W9-DGX-976

PERGAMON INTERNATIONAL LIBRARY
of Science, Technology, Engineering and Social Studies
*The 1000-volume original paperback library in aid of education,
industrial training and the enjoyment of leisure*
Publisher: Robert Maxwell, M.C.

Fluid Mechanics, Thermodynamics of Turbomachinery

THIRD EDITION in SI/METRIC UNITS

THE PERGAMON TEXTBOOK INSPECTION COPY SERVICE

An inspection copy of any book published in the Pergamon International Library will gladly be sent to academic staff without obligation for their consideration for course adoption or recommendation. Copies may be retained for a period of 60 days from receipt and returned if not suitable. When a particular title is adopted or recommended for adoption for class use and the recommendation results in a sale of 12 or more copies, the inspection copy may be retained with our compliments. The Publishers will be pleased to receive suggestions for revised editions and new titles to be published in this important International Library.

THERMODYNAMICS AND FLUID MECHANICS SERIES

General Editor: W. A. WOODS

Other Pergamon Titles of Interest

BENSON & WHITEHOUSE
Internal Combustion Engines (2 volumes)

DIXON
Worked Examples in Turbomachinery (Fluid Mechanics and Thermodynamics)

DUNN & REAY
Heat Pipes, 2nd Edition

GOSTELOW
Cascade Aerodynamics

HAYWOOD
Analysis of Engineering Cycles, 3rd Edition

JAFFEE
Rotor Forgings for Turbines and Generators

LAI *et al*
Introduction to Continuum Mechanics, Revised Edition

RAO
The Finite Element Method in Engineering

REAY & MACMICHAEL
Heat Pumps

Pergamon Related Journals
(*Free specimen copy gladly sent on request*)

International Journal of Heat and Mass Transfer
International Journal of Mechanical Sciences
International Journal of Multiphase Flow
Journal of Heat Recovery Systems
Letters in Heat and Mass Transfer
Mechanism and Machine Theory

Fluid Mechanics, Thermodynamics of Turbomachinery

S. L. DIXON, B.Eng., Ph.D., C.Eng., MI.Mech.E.

Lecturer in Mechanical Engineering at the University of Liverpool

THIRD EDITION in SI/METRIC UNITS

PERGAMON PRESS

OXFORD · NEW YORK · TORONTO · SYDNEY · FRANKFURT

U.K.	Pergamon Press Ltd., Headington Hill Hall, Oxford OX3 0BW, England
U.S.A.	Pergamon Press Inc., Maxwell House, Fairview Park, Elmsford, New York 10523, U.S.A.
CANADA	Pergamon Press Canada Ltd., Suite 104, 150 Consumers Road, Willowdale, Ontario M2J 1P9, Canada
AUSTRALIA	Pergamon Press (Aust.) Pty. Ltd., P.O. Box 544, Potts Point, N.S.W. 2011, Australia
FEDERAL REPUBLIC OF GERMANY	Pergamon Press GmbH, Hammerweg 6, D-6242 Kronberg, Federal Republic of Germany
JAPAN	Pergamon Press Ltd., 8th Floor, Matsuoka Central Building, 1-7-1 Nishishinjuku, Shinjuku-ku, Tokyo 160, Japan
BRAZIL	Pergamon Editora Ltda., Rua Eça de Queiros, 346, CEP 04011, São Paulo, Brazil
PEOPLE'S REPUBLIC OF CHINA	Pergamon Press, Qianmen Hotel, Beijing, People's Republic of China

Copyright © 1978 S. L. Dixon

First edition 1966
Second edition 1975
Third edition 1978
Reprinted 1979, 1982 (twice)
Reprinted with corrections 1984
Reprinted 1986

British Library Cataloguing in Publication Data

Dixon, Sydney Lawrence
Fluid mechanics, thermodynamics of
turbomachinery.—3rd ed. in SI/metric
units.—(Thermodynamics and fluid mechanics
series.)—(Pergamon international library).
1. Turbomachines—Fluid dynamics
2. Thermodynamics
I.Title   II. Series
621.8'11   TJ267   78–40103
ISBN 0–08–022721–X (hard cover)
ISBN 0–08–022722–8 (flexicover)

Printed in Great Britain by A. Wheaton & Co. Ltd., Exeter

Contents

v

4. Axial-flow Turbines: Two-dimensional Theory   93

5. Axial-flow Compressors, Pumps and Fans: Two-dimensional Analysis   120

6. Three-dimensional Flows in Axial Turbomachines 152

7. Centrifugal Pumps, Fans and Compressors 188

Preface to Third Edition

Several modifications have been incorporated into the text in the light of recent advances in some aspects of the subject. Further information on the interesting phenomenon of cavitation has been included and a new section on the optimum design of a pump inlet together with a worked example have been added which take into account recently published data on cavitation limitations. The chapter on *three-dimensional flows in axial turbomachines* has been extended; in particular the section concerning the *constant specific mass flow design* of a turbine nozzle has been clarified and now includes the flow equations for a following rotor row. Some minor alterations on the definition of blade shapes were needed so I have taken the opportunity of including a simplified version of the parabolic arc camber line as used for some low camber blading.

Despite careful proof reading a number of errors still managed to elude me in the second edition. I am most grateful to those readers who have detected errors and communicated with me about them.

In order to assist the reader I have (at last) added a list of symbols used in the text.

S.L.D.

Preface to Second Edition

THE first edition of *Fluid Mechanics, Thermodynamics of Turbomachinery* was very well received but has now been out of print for some time. In this revised and larger edition all dimensional quantities are given solely in SI units.* SI is now the only system of units used for teaching engineering in colleges, polytechnics and universities in the U.K. and it is fast becoming widely employed in many industrial concerns. Many other countries not previously metricated have recognised the advantages of the system and are beginning to use it. Recently, I was most interested (and relieved!) to see an announcement in *Transactions of the American Society of Mechanical Engineers* that in future all technical papers presented for publication would have to be in SI units.

The book follows the general lines of the first edition but now includes more worked solutions as these have been found to be most helpful to the student. Because the scope of the book has been broadened more problems having greater variety have been added to most of the chapters. If this is a source of dismay to the reader then he may be heartened to learn that a companion volume of worked solutions to these problems is due to be published in the near future.

I have extensively revised and enlarged the chapter on radial flow turbines as many new developments and ideas have emerged in recent years in various technical reports. The "spin-off" from the NASA space research programme has given a considerable impetus to the development and optimisation of the small radial turbine used in gas turbine power plants. This source has provided an important contribution to Chapter 8.

Several other chapters of the book have had new sections added to them either in the light of recent advances in knowledge or because of

---

*SI is the accepted symbol for the *Système International d'Unités* which is the modern form of the metric system agreed in 1960 by the General Conference on Weights and Measures (GCPM).

facts I have unearthed which have advanced my own understanding of the subject. I have added some theoretical and experimental information about diffusers and considered both American as well as British cascade correlations in more detail than I had attempted previously. More attention has been given to incompressible flow machines and this includes the precise definition of heads and efficiencies.

I am grateful for suggestions and useful comments which I have received on the first edition from former students, colleagues at Liverpool University and correspondents in several countries. In particular, I am indebted to Drs. D. J. Ryley and W. A. Woods for drawing my attention to several obscurities and minor errors which I had failed to notice. Comments and constructive criticism from correspondents in connection with the second edition would be most welcome. Last, but by no means least, I wish to thank Rosaleen Dixon for her general help with the book including the typing.

S. L. Dixon

List of Symbols

A	area
a	sonic velocity, position of maximum camber
b	passage width, maximum camber
C_f	tangential force coefficient
C_L, C_D	lift and drag coefficients
C_p	specific heat at constant pressure, pressure coefficient
C_v	specific heat at constant volume
c	absolute velocity
c_o	spouting velocity
D	drag force, diameter
D_{eq}	equivalent diffusion ratio
D_h	hydraulic mean diameter
E, e	energy, specific energy
f	acceleration
g	gravitational acceleration
H	head, blade height
H_s	net positive suction head (NPSH)
h	specific enthalpy
I	rothalpy
i	incidence angle
K, k	constants
L	lift force
l	blade chord length
M	Mach number
m	mass, molecular 'weight'
N	rotational speed
N_s	specific speed (rev)
N_{sp}	power specific speed (rev)
N_{ss}	suction specific speed (rev)
n	number of stages, polytropic index
p	pressure
Q	heat transfer, volume flow rate

q	dryness fraction
R	reaction, specific gas constant
Re	Reynolds number
R_H	reheat factor
R_o	Universal gas constant
r	radius
S	entropy
s	blade pitch, specific entropy
T	absolute temperature
t	time, thickness
U	blade speed, internal energy
u	specific internal energy
V, v	volume, specific volume
W	work transfer
ΔW	specific work transfer
w	relative velocity
X	axial force
x, y, z	coordinate directions
Y	tangential force
Y_p	profile loss coefficient
Z	number of blades, blade loading parameter
α	absolute flow angle
β	relative flow angle
Γ	circulation
γ	ratio of specific heats
δ	deviation angle
ε	fluid deflection angle, cooling effectiveness
ζ	enthalpy loss coefficient
η	efficiency
Θ	minimum opening
θ	blade camber angle, wake momentum thickness
λ	profile loss coefficient
μ	dynamic viscosity
υ	kinematic viscosity
ξ	blade stagger angle
ρ	density

σ	slip factor
σ_b	blade cavitation coefficient
τ	torque
ϕ	flow coefficient, velocity ratio
ψ	stage loading factor
Ω	speed of rotation (rad/s)
Ω_s	specific speed (rad)
Ω_{ss}	suction specific speed (rad)
ω	vorticity
$\bar{\omega}$	stagnation pressure loss coefficient

SUBSCRIPTS

av	average
c	compressor, critical
D	diffuser
e	exit
h	hydraulic, hub
i	inlet, impeller
id	ideal
is	isentropic
m	mean, meridional, mechanical
N	nozzle
n	normal component
o	stagnation property, overall
p	polytropic, constant pressure
R	reversible process, rotor
r	radial
rel	relative
s	isentropic, stall condition
ss	stage isentropic
t	turbine, tip, transverse
v	velocity
x, y, z	coordinate components
θ	tangential

SUPERSCRIPTS

˙	time rate of change
—	average
′	blade angle (as distinct from flow angle)
*	nominal condition

CHAPTER 1

Introduction:
Dimensional Analysis:
Similitude

If you have known one you have known all. (TERENCE, *Phormio.*)

DEFINITION OF A TURBOMACHINE

We classify as turbomachines all those devices in which energy is transferred either to, or from, a continuously flowing fluid by the *dynamic action* of one or more moving blade rows. The word *turbo* or *turbinis* is of Latin origin and implies that which spins or whirls around. Essentially, a rotating blade row or *rotor* changes the stagnation enthalpy of the fluid moving through it by either doing positive or negative work, depending upon the effect required of the machine. These enthalpy changes are intimately linked with the pressure changes occurring simultaneously in the fluid.

The definition of a turbomachine as stated above, is rather too general for the purposes of this book as it embraces *open* turbomachines such as propellers, windmills and unshrouded fans that influence an indeterminate quantity of fluid. The discussion, therefore, is limited to *enclosed* turbomachines in which a finite quantity of fluid passes through a *casing* in unit time. The subject of open turbomachines comes within the compass of general aerodynamics textbooks such as that of Glauert.[1]

Two main categories of turbomachine are identified: firstly, those that *absorb* power to increase the fluid pressure or head (ducted fans, compressors and pumps); secondly, those that *produce* power by expanding fluid to a lower pressure or head (hydraulic, steam and gas turbines). Figure 1.1 shows, in a simple diagrammatic form, a selection of the many different varieties of turbomachine encountered in practice.

1

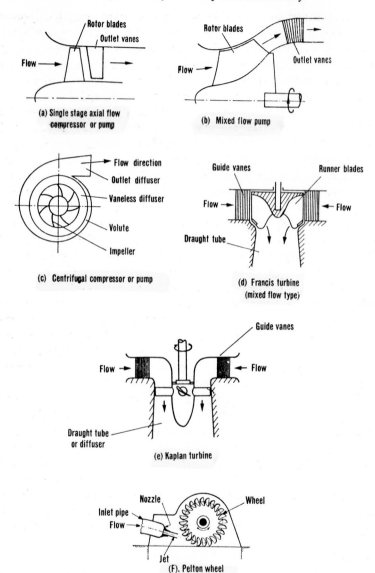

FIG. 1.1. Diagrammatic form of various types of turbomachine.

The reason that so many different types of either pump (compressor) or turbine are in use is because of the almost infinite range of service requirements. Generally speaking, for a given set of operating requirements there is one type of pump or turbine best suited to provide optimum conditions of operation. This point is discussed more fully in the section of this chapter concerned with specific speed.

Turbomachines are further categorised according to the nature of the flow path through the passages of the rotor. When the path of the *through-flow* is wholly or mainly parallel to the axis of rotation, the device is termed an *axial flow turbomachine* (e.g. Fig. 1.1(a) and (e)). When the path of the *through-flow* is wholly or mainly in a plane perpendicular to the rotation axis, the device is termed a *radial flow turbomachine* (e.g. Fig. 1.1(c)). More detailed sketches of radial flow machines are given in Figs. 7.1, 7.2 and 8.2. *Mixed flow turbomachines* are widely used. The term *mixed flow* in this context refers to the direction of the through-flow at rotor outlet when both radial and axial velocity components are present in significant amounts. Figure 1.1(b) shows a mixed flow pump and Fig. 1.1 (d) a mixed flow hydraulic turbine.

One further category should be mentioned. All turbomachines can be classified as either *impulse* or *reaction* machines according to whether pressure changes are absent or present respectively in the flow through the rotor. In an impulse machine all the pressure change takes place in one or more nozzles, the fluid being directed onto the rotor. The Pelton wheel, Fig. 1.1(f), is an example of an impulse turbine.

The main purpose of this book is to examine, through the laws of fluid mechanics and thermodynamics, the means by which the energy transfer is achieved in the chief types of turbomachine, together with the differing behaviour of individual types in operation. Methods of analysing the flow processes differ depending upon the geometrical configuration of the machine, on whether the fluid can be regarded as incompressible or not, and whether the machine absorbs or produces work. As far as possible, a unified treatment is adopted so that machines having similar configurations and function are considered together.

UNITS AND DIMENSIONS

The SI basic units used in fluid mechanics and thermodynamics are the *metre* (m), *kilogram* (kg), *second* (s) and *thermodynamic temperature*

(K). All other units used in this book are derived from these basic units. The *unit of force* is the NEWTON (N) defined as that force which, applied to a mass of 1 kilogram, gives it an acceleration of 1 metre per second squared. The recommended *unit of pressure* is the PASCAL (Pa) which is the pressure produced by a force of 1 newton uniformly distributed over an area of 1 square metre. Several other units of pressure are in widespread use, however, foremost of these being the BAR. Much basic data concerning properties of substances (steam and gas tables, charts, etc.) have been prepared in SI units with pressure given in bars and it is acknowledged that this alternative unit of pressure will continue to be used for some time as a matter of expediency. It is noted that 1 bar equals 10^5 Pa (i.e. 10^5 N/m²), roughly the pressure of the atmosphere at sea level, and is perhaps an inconveniently large unit for pressure in the field of turbomachinery anyway! In this book the convenient size of the kiloPascal (kPa) is found to be the most useful multiple of the recommended unit and is extensively used in most calculations and examples.

In SI the units of all forms of energy are the same as for work. The *unit of energy* is the JOULE (J) which is the work done when a force of 1 newton is displaced through a distance of 1 metre in the direction of the force, e.g. kinetic energy ($\frac{1}{2}mc^2$) has the dimensions kg × m²/s²; however, 1 kg = 1 N s²/m from the definition of the newton given above. Hence, the units of kinetic energy must be Nm = J upon substituting dimensions.

The WATT (W) is the *unit of power*; when 1 watt is applied for 1 second to a system the input of energy to that system is 1 joule (i.e. 1 J).

The HERTZ (Hz) is the number of repetitions of a regular occurrence in 1 second. Instead of writing c/s for cycles/sec, Hz is used instead.

The unit of thermodynamic temperature is the KELVIN (K), written without the ° sign, and is the fraction 1/273·16 of the thermodynamic temperature of the triple point of water. The DEGREE CELCIUS (°C) is equal to the unit kelvin. Zero on the Celcius scale is the temperature of the ice point (273·15 K). Specific heat capacity, or simply specific heat, is expressed as J/kg K or as J/kg°C.

Dynamic viscosity, dimensions $ML^{-1}T^{-1}$, has the SI units of Pascal seconds, i.e.

$$\frac{M}{LT} \equiv \frac{kg}{m.s} = \frac{N.s^2}{m.^2s} = Pa\,s.$$

Hydraulic engineers find it convenient to express pressure in terms of *head* of a liquid. The static pressure at any point in a liquid at rest is, relative to the pressure acting on the free surface, proportional to the vertical distance of the free surface above that point. The head H is simply the height of a column of the liquid which can be supported by this pressure. If ρ is the mass density (kg/m^3) and g the local gravitational acceleration (m/s^2), then the static pressure p (relative to atmospheric pressure) is $p = \rho g H$, where H is in metres and p is in Pascals (or N/m^2). This is left for the student to verify as a simple exercise.

There will, without doubt, be a few problems caused by the conversion to SI units. An outstanding problem to be recognised by the student or engineer is that many valuable papers or texts written prior to 1969 contain data in non-SI units. The booklet prepared by the National Physical Laboratory[6] provides the information necessary for the conversion of values expressed in the older British units of measurement to corresponding values in metric (SI) units. A brief summary of conversion factors between the more frequently used British units and SI units is given in an Appendix to the present volume. Further information of a supplementary nature is given in a Royal Society booklet[7] on internationally agreed definitions, names and symbols for units and on the rules for the expression of relations involving numbers between physical quantities and units.

DIMENSIONAL ANALYSIS AND PERFORMANCE LAWS

The widest comprehension of the general behaviour of all turbomachines is, without doubt, obtained from *dimensional analysis*. This is the formal procedure whereby the group of variables representing some physical situation is reduced into a smaller number of dimensionless groups. When the number of independent variables is not too great, dimensional analysis enables experimental relations between variables to be found with the greatest economy of effort. Dimensional analysis applied to turbomachines has two further important uses: (a) prediction of a prototype's performance from tests conducted on a scale model (similitude); (b) determination of the most suitable type of machine, on the basis of maximum efficiency, for a specified range of head, speed and flow rate. Several methods of constructing non-dimensional groups

have been described by Bradshaw[2] in a companion volume in this
series and Hunsaker and Rightmire[3] give a more detailed treatment of
the subject. It is assumed here that the basic techniques of forming
non-dimensional groups have already been acquired by the student.

Adopting the simple approach of elementary thermodynamics, an
imaginary envelope (called a *control surface*) of fixed shape, position
and orientation is drawn around the turbomachine (Fig. 1.2). Across

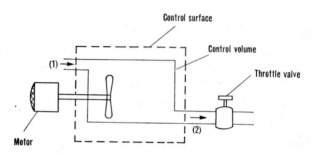

FIG. 1.2. Turbomachine considered as a control volume.

this boundary, fluid flows steadily, entering at station 1 and leaving at
station 2. As well as the flow of fluid there is a flow of work across the
control surface, transmitted by the shaft either to, or from, the machine.
For the present all details of the flow within the machine can be ignored
and only externally observed features such as shaft speed, flow rate,
torque and change in fluid properties across the machine need be considered. To be specific, let the turbomachine be a *pump* (although the
analysis could apply to other classes of turbomachine) driven by an
electric motor. The speed of rotation N, can be adjusted by altering the
current to the motor; the volume flow rate Q, can be *independently*
adjusted by means of a throttle valve. For fixed values of the set Q and
N, all other variables such as torque τ, head H, are thereby established.
The choice of Q and N as *control variables* is clearly arbitrary and any
other pair of independent variables such as τ and H could equally well
have been chosen. The important point to recognise is, that there are
for this pump, *two* control variables.

If the fluid flowing is changed for another of different density ρ,
and viscosity μ, the performance of the machine will be affected. Note,

also, that for a turbomachine handling compressible fluids, other *fluid properties* are important and are discussed later.

So far we have considered only one particular turbomachine, namely a pump of a given size. To extend the range of this discussion, the effect of the *geometric variables* on the performance must now be included. The size of machine is characterised by the impeller diameter D, and the shape can be expressed by a number of length ratios, l_1/D, l_2/D, etc.

INCOMPRESSIBLE FLUID ANALYSIS

The performance of a turbomachine can now be expressed in terms of the control variables, geometric variables and fluid properties. For the hydraulic pump it is convenient to regard the net energy transfer gH, the efficiency η, and power supplied P, as dependent variables and to write the three functional relationships as

$$gH = f_1\left(Q, N, D, \rho, \mu, \frac{l_1}{D}, \frac{l_2}{D}, \ldots\right), \tag{1.1a}$$

$$\eta = f_2\left(Q, N, D, \rho, \mu, \frac{l_1}{D}, \frac{l_2}{D}, \ldots\right), \tag{1.1b}$$

$$P = f_3\left(Q, N, D, \rho, \mu, \frac{l_1}{D}, \frac{l_2}{D}, \ldots\right), \tag{1.1c}$$

By the procedure of dimensional analysis using the three primary dimensions, mass, length and time, or alternatively, using three of the independent variables we can form the dimensionless groups. The latter, more direct procedure, requires that the variables selected, ρ, N, D, do not of themselves form a dimensionless group. The selection of ρ, N, D as common factors avoids the appearance of special fluid terms (e.g. μ, Q) in more than one group and allows gH, η and P to be made explicit. Hence the three relationships reduce to the following easily verified forms.

Energy transfer coefficient, sometimes called head coefficient

$$\psi = \frac{gH}{(ND)^2} = f_4\left(\frac{Q}{ND^3}, \frac{\rho ND^2}{\mu}, \frac{l_1}{D}, \frac{l_2}{D}, \ldots\right), \tag{1.2a}$$

$$\eta = f_5\left(\frac{Q}{ND^3}, \frac{\rho ND^2}{\mu}, \frac{l_1}{D}, \frac{l_2}{D}, \dots\right). \tag{1.2b}$$

Power coefficient

$$\hat{P} = \frac{P}{\rho N^3 D^5} = f_6\left(\frac{Q}{ND^3}, \frac{\rho ND^2}{\mu}, \frac{l_1}{D}, \frac{l_2}{D}, \dots\right). \tag{1.2c}$$

The non-dimensional group $Q/(ND^3)$ is a volumetric flow coefficient and $\rho ND^2/\mu$ is a form of Reynolds number, Re. In axial flow turbomachines, an alternative to $Q/(ND^3)$ which is frequently used is the velocity (or flow) coefficient $\phi = c_x/U$ where U is blade tip speed and c_x the average axial velocity. Since

$$Q = c_x \times \text{flow area} \propto c_x D^2$$

and $U \propto ND$.

then

$$\frac{Q}{ND^3} \propto \frac{c_x}{U}.$$

Because of the large number of independent groups of variables on the right-hand side of eqns. (1.2), those relationships are virtually worthless unless certain terms can be discarded. In a family of *geometrically similar* machines l_1/D, l_2/D are constant and may be eliminated forthwith. The kinematic viscosity, $\nu = \mu/\rho$ is very small in turbomachines handling water and, although speed, expressed by ND, is low the Reynolds number is correspondingly high. Experiments confirm that effects of Reynolds number on the performance are small and may be ignored in a first approximation. The functional relationships for geometrically similar hydraulic turbomachines are then,

$$\psi = f_4[Q/(ND^3)] \tag{1.3a}$$

$$\eta = f_5[Q/(ND^3)] \tag{1.3b}$$

$$\hat{P} = f_6[Q/(ND^3)]. \tag{1.3c}$$

This is as far as the reasoning of dimensional analysis alone can be taken; the actual *form* of the functions f_4, f_5 and f_6 must be ascertained by experiment.

One relation between ψ, ϕ, η and $\hat{P}$ may be immediately stated. For a pump the *net hydraulic power*, P_N equals ρQgH which is the minimum shaft power required in the absence of all losses. No real process of power conversion is free of losses and the actual shaft power P must be larger than P_N. We define pump efficiency (more precise definitions of efficiency are stated in Chapter 2) $\eta = P_N/P = \rho QgH/P$. Therefore

$$P = \frac{1}{\eta}\left(\frac{Q}{ND^3}\right)\frac{gH}{(ND)^2}\,\rho N^3 D^5. \tag{1.4}$$

Thus f_6 may be derived from f_4 and f_5 since $\hat{P} = \phi\psi/\eta$. For a turbine the net hydraulic power P_N supplied is greater than the actual shaft power delivered by the machine and the efficiency $\eta = P/P_N$. This can be rewritten as $\hat{P} = \eta\phi\psi$ by reasoning similar to the above considerations.

PERFORMANCE CHARACTERISTICS

The operating condition of a turbomachine will be *dynamically similar* at two different rotational speeds if all fluid velocities at *corresponding points* within the machine are in the same direction and proportional to the blade speed. If two points, one on each of two different head–flow characteristics, represent dynamically similar operation of the machine, then the non-dimensional groups of the variables involved, ignoring Reynolds number effects, may be expected to have the same numerical value for both points. On this basis, non-dimensional presentation of performance data has the important practical advantage of collapsing into virtually a single curve, results that would otherwise require a multiplicity of curves if plotted dimensionally.

Evidence in support of the foregoing assertion is provided in Fig. 1.3 which shows experimental results obtained by the author (at the University of Liverpool) on a simple centrifugal laboratory pump. Within the normal operating range of this pump, $0{\cdot}03 < Q/(ND^3) < 0{\cdot}06$, very little systematic scatter is apparent which might be associated with a Reynolds number effect, for the range of speeds $2500 \le N \le 5000$ rev/min. For smaller flows, $Q/(ND^3) < 0{\cdot}025$, the flow became unsteady and the manometer readings of uncertain accuracy but, nevertheless, dynamically similar conditions still appear to hold true. Examining the results at high flow rates one is struck by a marked systematic deviation away from the "single-curve" law at increasing speed. This effect is due

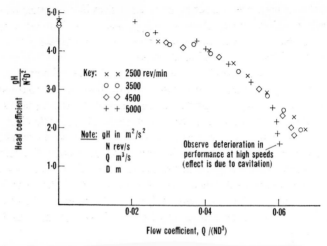

FIG. 1.3. Dimensionless head-volume characteristic of a centrifugal pump.

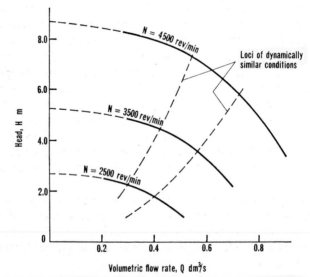

FIG. 1.4. Extrapolation of characteristic curves for dynamically similar conditions at $N = 3500$ rev/min.

to *cavitation*, a high speed phenomenon of hydraulic machines caused by the release of vapour bubbles at low pressures, which is discussed later in this chapter. It will be clear at this stage that under cavitating flow conditions, dynamical similarity is not possible.

The non-dimensional results shown in Fig. 1.3 have, of course, been obtained for a particular pump. They would also be approximately valid for a range of *different* pump sizes so long as all these pumps are geometrically similar and cavitation is absent. Thus, neglecting any change in performance due to change in Reynolds number, the dynamically similar results in Fig. 1.3 can be applied to predicting the dimensional performance of a given pump for a series of required speeds. Figure 1.4 shows such a dimensional presentation. It will be clear from the above discussion that the locus of dynamically similar points in the $H–Q$ field lies on a parabola since H varies as N^2 and Q varies as N.

VARIABLE GEOMETRY TURBOMACHINES

The efficiency of a fixed geometry machine, ignoring Reynolds number effects, is a unique function of flow coefficient. Such a dependence is shown by line (b) in Fig. 1.5. Clearly, off-design operation of such a machine is grossly inefficient and designers sometimes resort to a *variable geometry* machine in order to obtain a better match with changing flow conditions. Figure 1.6 shows a sectional sketch of a mixed-flow pump in which the impeller vane angles may be varied *during* pump operation. (A similar arrangement is used in Kaplan turbines, Fig. 1.1.) Movement of the vanes is implemented by cams driven from a servomotor. In some very large installations involving many thousands of kilowatts and where operating conditions fluctuate, sophisticated systems of control may incorporate an electronic computer.

The lines (a) and (c) in Fig. 1.5 show the efficiency curves at other blade settings. Each of these curves represents, in a sense, a different constant geometry machine. For such a variable geometry pump the desired operating line intersects the points of maximum efficiency of each of these curves.

Introducing the additional variable β into eqn. (1.3) to represent the setting of the vanes, we can write

$$\psi = f_1(\phi, \beta); \quad \eta = f_2(\phi, \beta). \tag{1.5}$$

Alternatively, with $\beta = f_3(\phi, \eta) = f_4(\phi, \psi)$, β can be eliminated to give a new functional dependence

$$\eta = f_5(\phi, \psi) = f_5\left(\frac{Q}{ND^3}, \frac{gH}{N^2 D^2}\right) \tag{1.6}$$

Thus, efficiency in a variable geometry pump is a function of both flow coefficient and energy transfer coefficient.

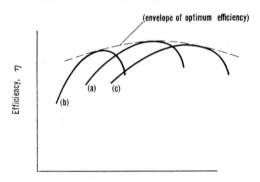

FIG. 1.5. Different efficiency curves for a given machine obtained with various blade settings.

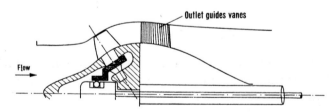

FIG. 1.6. Mixed-flow pump incorporating mechanism for adjusting blade setting.

SPECIFIC SPEED

The pump or hydraulic turbine designer is often faced with the basic problem of deciding what type of turbomachine will be the best choice for a given duty. He will usually be provided with preliminary design data such as head H, volume flow rate Q and rotational speed N when a pump design is being considered. For a turbine the power P, H and N would normally be specified. A non-dimensional parameter called *specific speed* N_s (sometimes termed the *shape number*) is used by the designer to aid his choice. This new parameter is derived from the non-dimensional groups defined in eqn. (1.3) in such a way that the characteristic diameter D of the turbomachine is eliminated. The value of N_s gives the designer a guide to the type of machine that will provide the normal requirement of high efficiency at the design condition.

For any one hydraulic turbomachine *with fixed geometry* there is a unique relationship between efficiency and flow coefficient if Reynolds number effects are negligible and cavitation absent. As is suggested by any one of the curves in Fig. 1.5, the efficiency rises to a maximum value as the flow coefficient is increased and then gradually falls with further increase in ϕ. This optimum efficiency $\eta = \eta_{max}$, is used to identify a unique value $\phi = \phi_1$ and corresponding unique values of $\psi = \psi_1$ and $\hat{P} = \hat{P}_1$. Thus,

$$\frac{Q}{ND^3} = \phi_1 = \text{constant}, \tag{1.7a}$$

$$\frac{gH}{N^2 D^2} = \psi_1 = \text{constant}, \tag{1.7b}$$

$$\frac{P}{\rho N^3 D^5} = \hat{P}_1 = \text{constant}. \tag{1.7c}$$

It is a simple matter to combine any pair of these expressions in such a way as to eliminate the diameter. For a pump the customary way of eliminating D is to divide $\phi_1^{\frac{1}{2}}$ by $\psi_1^{\frac{3}{4}}$. Thus

$$N_s = \frac{\phi_1^{\frac{1}{2}}}{\psi_1^{\frac{3}{4}}} = \frac{NQ^{\frac{1}{2}}}{(gH)^{\frac{3}{4}}}, \tag{1.8}$$

where N_s is called the *specific speed*. The term specific speed is justified to the extent that N_s is directly proportional to N. In the case of a turbine the *power specific speed* N_{sp}, is more useful and is defined by,

$$N_{sp} = \frac{\hat{P}_1{}^{\frac{1}{2}}}{\psi_1{}^{5/4}} = \frac{NP^{\frac{1}{2}}}{\rho^{\frac{1}{2}}(gH)^{5/4}}. \qquad (1.9)$$

Remembering that specific speed, as defined above, is at the point of maximum efficiency of a turbomachine, it becomes a parameter of great importance in selecting the type of machine required for a given

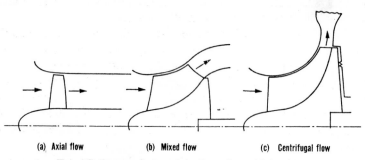

(a) Axial flow   (b) Mixed flow   (c) Centrifugal flow

FIG. 1.7. Range of pump impellers of equal inlet area.

duty. The maximum efficiency condition *replaces* the condition of geometric similarity, so that any alteration in specific speed implies that the machine design changes. Broadly speaking, each different class of machine has its optimum efficiency within its own fairly narrow range of specific speed.

For a pump, eqn. (1.8) indicates, for constant speed N, that N_s is increased by an increase in Q and decreased by an increase in H. From eqn. (1.7b) it is observed that H, at a constant speed N, is increased with impeller diameter D. Consequently, to increase N_s the entry area must be made large and/or the maximum impeller diameter small. Figure 1.7 shows a range of pump impellers varying from the axial-flow type, through mixed flow to a centrifugal- or radial-flow type. The size of each inlet is such that they all handle the same volume flow Q. Likewise, the head developed by each impeller (of different diameter D) is made equal by adjusting the speed of rotation N. Since Q and H are constant, then N_s varies with N alone. The most noticeable feature of this comparison is the large change in size with specific speed. Since a higher specific speed implies a smaller machine, for reasons of economy, it is desirable to select the *highest possible specific speed* consistent with good efficiency.

CAVITATION

In selecting a hydraulic turbomachine for a given head H and capacity Q, it is clear from the definition of specific speed, eqn. (1.8), that the highest possible value of N_s should be chosen because of the resulting reduction in size, weight and cost. On this basis a turbomachine could be made extremely small were it not for the corresponding increase in the fluid velocities. For machines handling liquids the lower limit of size is dictated by the phenomenon of *cavitation*.

Cavitation is the boiling of a liquid at normal temperature when the static pressure is made sufficiently low. It may occur at the entry to pumps or at the exit from hydraulic turbines in the vicinity of the moving blades. The dynamic action of the blades causes the static pressure to reduce locally in a region which is already normally below atmospheric pressure and cavitation can commence. The phenomenon is accentuated by the presence of dissolved gases which are released with a reduction in pressure.

For the purpose of illustration consider a centrifugal pump operating at constant speed and capacity. By steadily reducing the inlet pressure head a point is reached when streams of small vapour bubbles appear within the liquid and close to solid surfaces. This is called *cavitation inception* and commences in the regions of lowest pressure. These bubbles are swept into regions of higher pressure where they collapse. This condensation occurs suddenly, the liquid surrounding the bubbles either hitting the walls or adjacent liquid. The pressure wave produced by bubble collapse (with a magnitude of the order 400 MPa) momentarily raises the pressure level in the vicinity and the action ceases. The cycle then repeats itself and the frequency may be as high as 25 kHz[4]. The repeated action of bubbles collapsing near solid surfaces leads to the well-known cavitation erosion.

The collapse of vapour cavities generates noise over a wide range of frequencies—up to 1 MHz has been measured,[9] i.e. so-called "white noise". Apparently it is the collapsing smaller bubbles which cause the higher frequency noise and the larger cavities the lower frequency noise. Noise measurement can be used as a means of detecting cavitation.[10] Pearsall has shown experimentally[11] that there is a relationship between cavitation noise levels and erosion damage on cylinders and concludes that a technique could be developed for predicting the occurrence of erosion.

Up to this point no detectable deterioration in performance has occurred. However, with further reduction in inlet pressure, the bubbles increase both in size and number, coalescing into pockets of vapour which affects the whole field of flow. This growth of vapour cavities is usually accompanied by a sharp drop in pump performance as shown conclusively in Fig. 1.3 (for the 5000 rev/min test data). It may seem surprising to learn that with this large change in bubble size, the solid surfaces are much less likely to be damaged than at inception of cavitation. The avoidance of cavitation inception in conventionally designed machines can be regarded as one of the essential tasks of both pump and turbine designers. However, in certain recent specialised applications pumps have been designed to operate under *supercavitating* conditions. Under these conditions large size vapour bubbles are formed but, bubble collapse takes place *downstream* of the impeller blades. An example of the specialised application of a supercavitating pump is the fuel pumps of rocket engines for space vehicles where size and mass must be kept low at all costs. Pearsall[8] has shown that the supercavitating principle is most suitable for axial flow pumps of high specific speed and has suggested a design technique using methods similar to those employed for conventional pumps.

CAVITATION LIMITS

In theory cavitation commences in a liquid when the static pressure is reduced to the vapour pressure corresponding to the liquid's temperature. However, in practice, the physical state of the liquid will determine the pressure at which cavitation starts.[9] Dissolved gases come out of solution as the pressure is reduced forming gas cavities at pressures in excess of the vapour pressure. Vapour cavitation requires the presence of nuclei —submicroscopic gas bubbles or solid non-wetted particles—in sufficient numbers. It is an interesting fact that in the absence of such nuclei a liquid can withstand negative pressures (i.e. tensile stresses) of the order of tens of atmosphere. Special pre-treatment (i.e. rigorous filtration and pre-pressurization) of the liquid is required to obtain this state. In general the liquids flowing through turbomachines will contain some dust and dissolved gases and under these conditions negative pressures do not arise.

A useful parameter is the available suction head at entry to a pump or at exit from a turbine. This is usually referred to as the *net positive suction head*, NPSH, defined as

$$H_s = (p_o - p_v)/(\rho g) \qquad (1.10)$$

where p_o and p_v are the absolute stagnation and vapour pressures, respectively, at pump inlet or at turbine outlet.

To take into account the effects of cavitation, the performance laws of a hydraulic turbomachine should include the additional independent variable H_s. Ignoring the effects of Reynolds number, the performance laws of a constant geometry hydraulic turbomachine are then dependent on two groups of variable. Thus, the efficiency,

$$\eta = f(\phi, N_{ss}) \qquad (1.11)$$

where the *suction specific speed* $N_{ss} = NQ^{\frac{1}{2}}/(gH_s)^{\frac{3}{4}}$, determines the effect of cavitation, and $\phi = Q/(ND^3)$, as before.

It is known from experiment that cavitation inception occurs for an almost constant value of N_{ss} for all pumps (and, separately, for all turbines) designed to resist cavitation. This is because the blade sections at the inlet to these pumps are broadly similar (likewise, the exit blade sections of turbines are similar) and it is the *shape* of the low pressure passages which influences the onset of cavitation.

Using the alternative definition of suction specific speed $\Omega_{ss} = \Omega Q^{\frac{1}{2}}/(gH_s)^{\frac{3}{4}}$, where Ω is the rotational speed in rad/s, Q is the volume flow in m³/s and gH_s is in m²/s², it has been shown empirically[5] that

$$\Omega_{ss} \simeq 3\cdot0 \text{ (rad)} \qquad (1.12a)$$

for pumps, and

$$\Omega_{ss} \simeq 4\cdot0 \text{ (rad)} \qquad (1.12b)$$

for turbines.

COMPRESSIBLE FLUID ANALYSIS

The application of dimensional analysis to compressible fluids increases, not unexpectedly, the complexity of the functional relationships obtained in comparison with those already found for incompressible fluids. Even if the fluid is regarded as a perfect gas, in addition to the previously used fluid properties, two further characteristics are required; these are a_{01}, the stagnation speed of sound at entry to the machine and γ, the ratio of specific heats C_p/C_v. In the following analysis the compressible fluids under discussion are either perfect gases, or else, dry vapours approximating in behaviour to a perfect gas.

Another choice of variables is usually preferred when appreciable density changes occur across the machine. Instead of volume flow rate Q, the mass flow rate $\dot{m}$ is used; likewise for the head change H, the isentropic *stagnation enthalpy* change Δh_{0s} is employed.

The choice of this last variable is a significant one for, in an ideal and adiabatic process, Δh_{0s} is equal to the work done by unit mass of fluid. This will be discussed still further in Chapter 2. Since heat transfer from the casings of turbomachines is, in general, of negligible magnitude compared with the flux of energy through the machine, temperature on its own may be safely excluded as a fluid variable. However, temperature is an easily observable characteristic and, for a perfect gas, can be easily introduced at the last by means of the equation of state, $p/\rho = RT$, where $R = R_0/m = C_p - C_v$, m being the molecular weight of the gas and $R_0 = 8.314$ kJ/(kg mol K) is the *Universal gas constant*.

The performance parameters Δh_{0s}, η and P for a turbomachine handling a compressible flow, are expressed functionally as:

$$\Delta h_{0s}, \eta, P = f(\mu, N, D, \dot{m}, \rho_{01}, a_{01}, \gamma). \tag{1.13}$$

Because ρ_0 and a_0 change through a turbomachine, values of these fluid variables are selected at inlet, denoted by subscript 1. Equation (1.13) expresses *three* separate functional relationships, each of which consists of eight variables. Again, selecting ρ_{01}, N, D as common factors each of these three relationships may be reduced to five dimensionless groups,

$$\frac{\Delta h_{0s}}{N^2 D^2}, \eta, \frac{P}{\rho_{01} N^3 D^5} = f\left\{\frac{\dot{m}}{\rho_{01} N D^3}, \frac{\rho_{01} N D^2}{\mu}, \frac{ND}{a_{01}}, \gamma\right\}. \tag{1.14}$$

Alternatively, the flow coefficient $\phi = \dot{m}/(\rho_{01} N D^3)$ can be written as $\phi = \dot{m}/(\rho_{01} a_{01} D^2)$. As ND is proportional to blade speed, the group ND/a_{01} is regarded as a *blade Mach number*.

For a machine handling a perfect gas a different set of functional relationships is often more useful. These may be found either by selecting the appropriate variables for a perfect gas and working through again from first principles or, by means of some rather straightforward transformations, rewriting eqn. (1.14) to give more suitable groups. The latter procedure is preferred here as it provides a useful exercise.

As a concrete example consider an adiabatic compressor handling a perfect gas. The isentropic stagnation enthalpy rise can now be written $C_p(T_{02s}-T_{01})$ for the perfect gas. This compression process is illustrated in Fig. 1.8a where the stagnation state point changes at constant

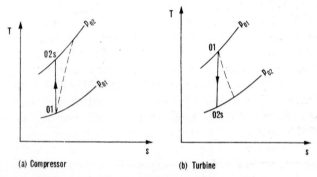

(a) Compressor   (b) Turbine

FIG. 1.8. The ideal adiabatic change in stagnation conditions across a turbomachine.

entropy between the stagnation pressures p_{01} and p_{02}. The equivalent process for a turbine is shown in Fig. 1.8b. Using the adiabatic isentropic relationship $p/\rho^\gamma = \text{constant}$, together with $p/\rho = RT$, the expression

$$\frac{T_{02s}}{T_{01}} = \left(\frac{p_{02}}{p_{01}}\right)^{(\gamma-1)/\gamma}$$

is obtained. Hence $\Delta h_{0s} = C_p T_{01}[(p_{02}/p_{01})^{(\gamma-1)/\gamma} - 1]$. Since $C_p = \gamma R/(\gamma-1)$ and $a_{01}^2 = \gamma R T_{01}$, then

$$\Delta h_{0s}/a_{01}^2 \propto f(p_{02}/p_{01}).$$

The flow coefficient can now be more conveniently expressed as

$$\frac{\dot{m}}{\rho_{01}a_{01}D^2} = \frac{\dot{m}RT_{01}}{p_{01}\sqrt{(\gamma RT_{01})D^2}} = \frac{\dot{m}\sqrt{(RT_{01})}}{D^2 p_{01}\sqrt{\gamma}}.$$

As $\dot{m} \equiv \rho_{01} D^2(ND)$, the power coefficient may be written

$$\hat{P} = \frac{P}{\rho_{01} N^3 D^5} = \frac{\dot{m} C_p \Delta T_0}{\{\rho_{01} D^2(ND)\}(ND)^2} = \frac{C_p \Delta T_0}{(ND)^2} \equiv \frac{\Delta T_0}{T_{01}}.$$

Collecting together all these newly formed non-dimensional groups and inserting them in eqn. (1.14) gives

$$\frac{p_{02}}{p_{01}}, \eta, \frac{\Delta T_0}{T_{01}} = f\left\{\frac{\dot{m}\sqrt{(RT_{01})}}{D^2 p_{01}}, \frac{ND,}{\sqrt{(RT_{01})}}, Re, \gamma\right\}. \tag{1.15}$$

The justification for dropping γ from a number of these groups is simply that it already appears separately as an independent variable.

For a machine of a given size and handling only a single gas, it is customary, in practice, to delete γ, R and D from eqn. (1.15). If, in addition, the machine operates at high Reynolds numbers (or over a small speed range), Re can also be dropped. Under these conditions eqn. (1.15) becomes

$$\frac{p_{02}}{p_{01}}, \eta, \frac{\Delta T_0}{T_{01}} = f\left\{\frac{\dot{m}\sqrt{T_{01}}}{p_{01}}, \frac{N}{\sqrt{T_{01}}}\right\}. \tag{1.16}$$

Note that by omitting the diameter D and gas constant R, the independent variables in eqn. (1.16) are no longer dimensionless.

Figures 1.9 and 1.10 represent typical performance maps obtained from compressor and turbine test results. In both figures the pressure ratio across the whole machine is plotted as a function of $\dot{m}(\sqrt{T_{01}})/p_{01}$ for fixed values of $N/(\sqrt{T_{01}})$, this being a customary method of presentation. Notice that for both machines subscript 1 is used to denote conditions at inlet. One of the most striking features of these performance characteristics is the rather weak dependence of the turbine performance upon $N/\sqrt{T_{01}}$ contrasting with the strong dependence shown by the compressor on this parameter.

For the compressor, efficient operation at constant $N/\sqrt{T_{01}}$ lies to the right of the line marked "*surge*". A discussion of the phenomenon of surge is included in Chapter 5; in brief, for multistage compressors it commences approximately at the point (for constant $N/\sqrt{T_{01}}$) where the pressure ratio flattens out to its maximum value. The surge line denotes the limit of *stable operation* of a compressor, unstable operation

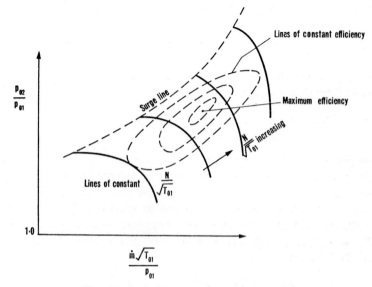

Lines of constant efficiency

Maximum efficiency

$\dfrac{N}{\sqrt{T_{01}}}$ increasing

Surge line

Lines of constant $\dfrac{N}{\sqrt{T_{01}}}$

$\dfrac{p_{02}}{p_{01}}$

1·0

$\dfrac{\dot{m}\sqrt{T_{01}}}{p_{01}}$

FIG. 1.9. Overall characteristic of a compressor.

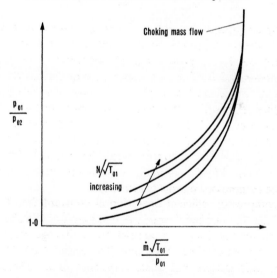

Choking mass flow

$\dfrac{p_{01}}{p_{02}}$

$N/\sqrt{T_{01}}$ increasing

1·0

$\dfrac{\dot{m}\sqrt{T_{01}}}{p_{01}}$

FIG. 1.10. Overall characteristic of a turbine.

being characterised by a severe oscillation of the mass flow rate through the machine. The choked regions of both the compressor and turbine characteristics may be recognised by the vertical portions of the constant speed lines. No further increase in $\dot{m}(\sqrt{T_{01}})/p_{01}$ is possible since the Mach number across some section of the machine has reached unity and the flow is said to be *choked*.

REFERENCES

1. GLAUERT, H., *The Elements of Aerofoil and Airscrew Theory*. Cambridge University Press (1959).
2. BRADSHAW, P., *Experimental Fluid Mechanics*. Pergamon Press, Oxford (2nd ed.) (1970).
3. HUNSAKER, J. C. and RIGHTMIRE, B. C., *Engineering Applications of Fluid Mechanics*. McGraw-Hill, New York (1947).
4. SHEPHERD, D. G., *Principles of Turbomachinery*. Macmillan, New York (1956).
5. WISLICENUS, G. F., *Fluid Mechanics of Turbomachinery*, McGraw-Hill, New York (1947).
6. ANDERTON, P. and BIGG, P. H., *Changing to the Metric System*. National Physical Laboratory, London, H.M.S.O. (1972).
7. SYMBOLS COMMITTEE OF THE ROYAL SOCIETY, *Quantities, Units and Symbols*. The Royal Society (1971).
8. PEARSALL, I. S., The design and performance of supercavitating pumps. *Proc. Symposium on Pump Design, Testing and Operation at N.E.L. Glasgow*. H.M.S.O. (1966).
9. PEARSALL, I. S., *Cavitation*. M&B Monograph ME/10. Mills & Boon (1972).
10. PEARSALL, I. S., Acoustic detection of cavitation. Symposium on Vibrations in Hydraulic Pumps and Turbines. *Proc. Instn. Mech. Engrs., London*, **181**, Pt. 3A (1966–7).
11. PEARSALL, I. S. and McNULTY, P. J., Comparison of cavitation noise with erosion. Cavitation Forum, 6-7, *Am. Soc. Mech. Engrs.* (1968).

PROBLEMS

1. A fan operating at 1750 rev/min at a volume flow rate of 4·25 m³/s develops a head of 153 mm measured on a water-filled U-tube manometer. It is required to build a larger, geometrically similar fan which will deliver the same head at the same efficiency as the existing fan, but at a speed of 1440 rev/min. Calculate the volume flow rate of the larger fan.

2. An axial flow fan 1·83 m diameter is designed to run at a speed of 1400 rev/min with an average axial air velocity of 12·2 m/s. A quarter scale model has been built to obtain a check on the design and the rotational speed of the model fan is 4200 rev/min. Determine the axial air velocity of the model so that dynamical similarity with the full-scale fan is preserved. The effects of Reynolds number change may be neglected.

A sufficiently large pressure vessel becomes available in which the complete model can be placed and tested under conditions of complete similarity. The viscosity of the air is independent of pressure and the temperature is maintained constant. At what pressure must the model be tested?

3. A water turbine is to be designed to produce 27 MW when running at 93·7 rev/min under a head of 16·5 m. A model turbine with an output of 37·5 kW is to be tested under dynamically similar conditions with a head of 4·9 m. Calculate the model speed and scale ratio. Assuming a model efficiency of 88%, estimate the volume flow rate through the model.

It is estimated that the force on the thrust bearing of the full-size machine will be 7·0 GN. For what thrust must the model bearing be designed?

4. Derive the non-dimensional groups that are normally used in the testing of gas turbines and compressors.

A compressor has been designed for normal atmospheric conditions (101·3 kPa and 15°C). In order to economise on the power required it is being tested with a throttle in the entry duct to reduce the entry pressure. The characteristic curve for its normal design speed of 4000 rev/min is being obtained on a day when the ambient temperature is 20°C. At what speed should the compressor be run? At the point of the characteristic curve at which the mass flow would normally be 58 kg/s the entry pressure is 55 kPa. Calculate the actual rate of mass flow during the test.

Describe, with the aid of sketches, the relationship between geometry and specific speed for pumps.

CHAPTER 2

Basic Thermodynamics, Fluid Mechanics: Definitions of Efficiency

Take your choice of those that can best aid your action. (SHAKESPEARE, *Coriolanus.*)

THIS chapter summarises the basic physical laws of fluid mechanics and thermodynamics, developing them into a form suitable for the study of turbomachines. Following this, some of the more important and commonly used expressions for the efficiency of compression and expansion flow processes are given.

The laws discussed are:

    (1)   the continuity equation;
    (2)   the First Law of Thermodynamics;
    (3)   Newton's Second Law of Motion;
    (4)   the Second Law of Thermodynamics.

The first three of these items are comprehensively dealt with in Volume 1 of this series[1] and the fourth item in Volume 2,[2] so that much of the elementary discussion and analysis of these laws need not be repeated here. It should be remembered, however, that these laws are completely general and that they are independent of the nature of the fluid flowing or whether the fluid is compressible or incompressible.

THE EQUATION OF CONTINUITY

Consider the flow of a fluid with density ρ, through the element of area dA, during the time interval dt. Referring to Fig. 2.1, if c is the stream velocity the elementary mass is $dm = \rho c \, dt \, dA \cos \theta$, where θ is the angle subtended by the normal of the area element to the stream direction. The velocity component perpendicular to the area dA is $c_n = c \cos \theta$ and so $dm = \rho c_n dA dt$. The elementary rate of mass flow is therefore

$$dm = \frac{dm}{dt} = \rho c_n dA. \tag{2.1}$$

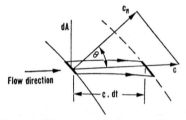

Fig. 2.1. Flow across an element of area.

Most analyses in this book are limited to one-dimensional steady flows where the velocity and density are regarded as constant across each section of a duct or passage. If A_1 and A_2 are the flow areas at stations 1 and 2 along a passage respectively, then

$$\dot{m} = \rho_1 c_{n1} A_1 = \rho_2 c_{n2} A_2 = \rho c_n A, \tag{2.2}$$

since there is no accumulation of fluid within the control volume.

THE FIRST LAW OF THERMODYNAMICS—INTERNAL ENERGY

The First Law states that if a system is taken through a complete cycle during which heat is supplied and work is done, then

$$\oint (dQ - dW) = 0, \tag{2.3}$$

where $\oint dQ$ represents the heat supplied to the system during the cycle and $\oint dW$ the work done by the system during the cycle. The units of heat and work in eqn. (2.3) are taken to be the same.

During a change of state from 1 to 2, there is a change in the property internal energy

$$E_2 - E_1 = \int_1^2 (dQ - dW). \tag{2.4}$$

For an infinitesimal change of state

$$dE = dQ - dW. \tag{2.4a}$$

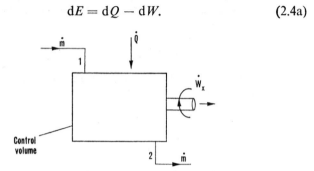

FIG. 2.2. Control volume showing sign convention for heat and work transfers.

The steady flow energy equation

It is shown in a number of textbooks[1,3,4] how the First Law is applied to the steady flow of fluid through a control volume so as to obtain the steady flow energy equation. It is unprofitable to reproduce this proof here and only the final result is quoted. Figure 2.2 shows a control volume representing a turbomachine, through which fluid passes at a steady rate of mass flow $\dot{m}$, entering at position 1 and leaving at position 2. Energy is transferred from the fluid to the blades of the turbomachine, positive work being done (via the shaft) at the rate $\dot{W}_x$. In the general case positive heat transfer takes place at the rate $\dot{Q}$, *from* the surroundings *to* the control volume. Thus, with this sign convention the steady flow energy equation is

$$\dot{Q} - \dot{W}_x = \dot{m}[(h_2 - h_1) + \tfrac{1}{2}(c_2^2 - c_1^2) + g(z_2 - z_1)], \tag{2.5}$$

where h is the specific enthalpy, $\tfrac{1}{2}c^2$ the kinetic energy per unit mass and gz the potential energy per unit mass.

Apart from hydraulic machines, the contribution of the last term in eqn. (2.5) is small and usually ignored. Defining stagnation enthalpy by $h_0 = h + \frac{1}{2}c^2$ and assuming $g(z_2 - z_1)$ is negligible, eqn. (2.5) becomes

$$\dot{Q} - \dot{W}_x = \dot{m}(h_{02} - h_{01}). \tag{2.6}$$

Most turbomachinery flow processes are adiabatic (or very nearly so) and it is permissible to write $\dot{Q} = 0$. For work producing machines (turbines) $\dot{W}_x > 0$, so that

$$\dot{W}_x = \dot{W}_t = \dot{m}(h_{01} - h_{02}). \tag{2.7}$$

For work absorbing machines (compressors) $\dot{W}_x < 0$, so that it is more convenient to write

$$W_c = -\dot{W}_x = \dot{m}(h_{02} - h_{01}). \tag{2.8}$$

NEWTON'S SECOND LAW OF MOTION

The fundamental principle of dynamics is Newton's Second Law of Motion. As this law is a vector relationship it requires a statement of the direction along which it is applied. Considering a system of mass m, the algebraic sum of the body and surface forces acting on m along some arbitrary direction x equals the time rate of change of the total x-momentum of the system, or

$$\Sigma F_x = m \frac{dc_x}{dt}. \tag{2.9}$$

For a control volume with fluid entering with uniform velocity c_{x1} and leaving with uniform velocity c_{x2}, then

$$\Sigma F_x = \dot{m}(c_{x2} - c_{x1}). \tag{2.9a}$$

Equation (2.9a) is the one-dimensional form of the steady flow momentum equation which states that the algebraic sum of the body forces on the matter within the control volume, plus the forces acting on the control surface, both in the x direction, equals the net outgoing flux of momentum in the x direction. This equation is important in the study of turbomachines as it enables the forces acting on blades and bearings to be determined.

Euler's equation of motion

It may be shown[1] for the steady flow of fluid through an elementary control volume that, in the absence of all shear forces, the relation

$$\frac{1}{\rho}\, dp + c\,dc + g\,dz = 0 \tag{2.10}$$

is obtained. This is Euler's equation of motion for one-dimensional flow and is derived from Newton's second law. By shear forces being absent we mean there is neither friction nor shaft work. However, it is not necessary that heat transfer should also be absent.

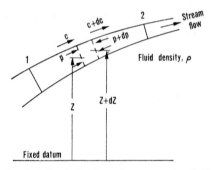

FIG. 2.3. Control volume in a streaming fluid.

Bernoulli's equation

The one-dimensional form of Euler's equation applies to a control volume whose thickness is infinitesimal in the stream direction (Fig. 2.3). Integrating this equation in the stream direction we obtain

$$\int_{1}^{2}\frac{1}{\rho}\, dp + \tfrac{1}{2}(c_2^2 - c_1^2) + g(z_2 - z_1) = 0 \tag{2.10a}$$

which is Bernoulli's equation. For an incompressible fluid, ρ is constant and eqn. (2.10a) becomes

$$\frac{1}{\rho}(p_{02} - p_{01}) + g(z_2 - z_1) = 0, \tag{2.10b}$$

where stagnation pressure is $p_0 = p + \tfrac{1}{2}\rho c^2$.

When dealing with hydraulic turbomachines, the term *head H* occurs frequently and describes the quantity $z + p_0/(\rho g)$. Thus eqn. (2.10b) becomes

$$H_2 - H_1 = 0. \qquad (2.10c)$$

If the fluid is a gas or vapour, the change in gravitational potential is generally negligible and eqn. (2.10a) is then

$$\int_1^2 \frac{1}{\rho}\,dp + \tfrac{1}{2}(c_2^2 - c_1^2) = 0. \qquad (2.10d)$$

Now, if the gas or vapour is subject to only a small pressure change the fluid density is sensibly constant and

$$p_{02} = p_{01} = p_0, \qquad (2.10e)$$

i.e. the stagnation pressure is constant (this is also true for a *compressible isentropic process*).

Moment of momentum

In dynamics much useful information is obtained by employing Newton's second law in the form where it applies to the moments of forces. This form is of central importance in the analysis of the energy transfer process in turbomachines.

For a system of mass m, the vector sum of the moments of all external forces acting on the system about some arbitrary axis $A-A$ fixed in space is equal to the time rate of change of angular momentum of the system about that axis, i.e.

$$\tau_A = m\,\frac{d}{dt}\,(rc_\theta), \qquad (2.11)$$

where r is distance of the mass centre from the axis of rotation measured along the normal to the axis and c_θ the velocity component mutually perpendicular to both the axis and radius vector r.

For a control volume the *law of moment of momentum* can be obtained.[1] Figure 2.4 shows the control volume enclosing the rotor of

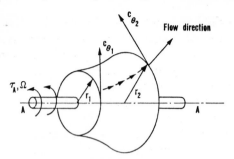

FIG. 2.4. Control volume for a generalised turbomachine.

a generalised turbomachine. Swirling fluid enters the control volume at radius r_1 with tangential velocity $c_{\theta 1}$ and leaves at radius r_2 with tangential velocity $c_{\theta 2}$. For one-dimensional steady flow

$$\tau_A = \dot{m}(r_2 c_{\theta 2} - r_1 c_{\theta 1}) \tag{2.11a}$$

which states that, the sum of the moments of the external forces acting on fluid temporarily occupying the control volume is equal to the net time rate of efflux of angular momentum from the control volume.

For a pump or compressor rotor running at angular velocity Ω, the rate at which the rotor does work on the fluid is

$$\tau_A \Omega = \dot{m}(U_2 c_{\theta 2} - U_1 c_{\theta 1}), \tag{2.12}$$

where the blade speed $U = \Omega r$.

Thus the work done on the fluid per unit mass or specific work, is

$$\Delta W_c = \frac{\dot{W}_c}{\dot{m}} = \frac{\tau_A \Omega}{\dot{m}} = U_2 c_{\theta 2} - U_1 c_{\theta 1} > 0. \tag{2.12a}$$

This equation is referred to as *Euler's pump equation*.

For a turbine the fluid does work *on* the rotor and the sign for work is then reversed. Thus, the specific work is

$$\Delta W_t = \frac{\dot{W}_t}{\dot{m}} = U_1 c_{\theta 1} - U_2 c_{\theta 2} > 0. \tag{2.12b}$$

Equation (2.12b) will be referred to as *Euler's turbine equation*.

THE SECOND LAW OF THERMODYNAMICS—ENTROPY

In order to specify the degree of imperfection of actual flow processes in turbomachines, an ideal process is required for comparison. The Second Law of Thermodynamics, rigorously developed in Volume 2 of this series[2] and in most modern textbooks on thermodynamics (for example, refs. 3 and 4), enables the concept of entropy to be introduced and ideal processes to be defined.

An important corollary of the Second Law known as the *Inequality of Clausius* states, for a system passing through a cycle involving heat exchanges, that

$$\oint \frac{\mathrm{d}Q}{T} \leqq 0, \tag{2.13}$$

where $\mathrm{d}Q$ is an element of heat transferred to the system at an absolute temperature T. If all the processes in the cycle are reversible then $\mathrm{d}Q = \mathrm{d}Q_R$ and the equality in eqn. (2.13) holds true, i.e.

$$\oint \frac{\mathrm{d}Q_R}{T} = 0. \tag{2.13a}$$

The property called entropy, for a finite change of state, is then defined as

$$S_2 - S_1 = \int_1^2 \frac{\mathrm{d}Q_R}{T}. \tag{2.14}$$

For an incremental change of state

$$\mathrm{d}S = m\,\mathrm{d}s = \frac{\mathrm{d}Q_R}{T}, \tag{2.14a}$$

where m is the mass of the system.

With steady one-dimensional flow through a control volume in which the fluid experiences a change of state from condition 1 at entry to 2 at exit,

$$\int_1^2 \frac{\mathrm{d}\dot{Q}}{T} \leqq \dot{m}(s_2 - s_1). \tag{2.15}$$

If the process is adiabatic, $d\dot{Q} = 0$, so that

$$s_2 \geqq s_1. \tag{2.16}$$

If the process is *reversible* as well,

$$s_2 = s_1. \tag{2.16a}$$

Thus, for a flow which is adiabatic, the ideal process will be one in which the entropy remains unchanged during the process (the condition of *isentropy*).

Several important expressions can be obtained using the above definition of entropy. For a system of mass m undergoing a reversible process $dQ = dQ_R = mTds$ and $dW = dW_R = mpdv$. In the absence of motion, gravity and other effects the First Law of Thermodynamics, eqn. (2.4a) becomes

$$Tds = du + pdv. \tag{2.17}$$

With $h = u + pv$ then $dh = du + pdv + vdp$ and eqn. (2.17) then gives

$$Tds = dh - vdp. \tag{2.18}$$

DEFINITIONS OF EFFICIENCY

A large number of efficiency definitions are included in the literature of turbomachines and most workers in this field would agree there are too many. In this book only those considered to be important and useful are included.

Efficiency of turbines

Turbines are designed to convert the available energy in a flowing fluid into useful mechanical work delivered at the coupling of the output shaft. The efficiency of this process, the *overall efficiency* η_0, is a performance factor of considerable interest to both designer and user of the turbine. Thus,

$$\eta_0 = \frac{\text{mechanical energy available at coupling of output shaft in unit time}}{\text{maximum energy difference possible for the fluid in unit time}}.$$

Mechanical energy losses occur between the turbine rotor and the output shaft coupling as a result of the work done against friction at the bearings, glands, etc. The magnitude of this loss as a fraction of the total energy transferred to the rotor is difficult to estimate as it varies with the size and individual design of turbomachine. For small machines (several kilowatts) it may amount to 5% or more, but for medium and large machines this loss ratio may become as little as 1%. A detailed consideration of the mechanical losses in turbomachines is beyond the scope of this book and is not pursued further.

The *adiabatic efficiency* η_t or *hydraulic efficiency* η_h for a turbine is, in broad terms,

$$\eta_t \text{ (or } \eta_h) = \frac{\text{mechanical energy supplied to the rotor in unit time}}{\text{maximum energy difference possible for the fluid in unit time}}.$$

Comparing the above definitions it is easily deduced that the *mechanical efficiency* η_m, which is simply the ratio of shaft power to rotor power, is

$$\eta_m = \eta_0/\eta_t \quad \text{(or } \eta_0/\eta_h).$$

In the following paragraphs the various definitions of hydraulic and adiabatic efficiency are discussed in more detail.

For an incremental change of state through a turbomachine the steady flow energy equation, eqn. (2.5), can be written

$$d\dot{Q} - d\dot{W}_x = \dot{m}[dh + \tfrac{1}{2}d(c^2) + gdz].$$

From the Second Law of Thermodynamics

$$d\dot{Q} \leqslant \dot{m}\,Tds = \dot{m}\left(dh - \frac{1}{\rho}\,dp\right).$$

Eliminating $d\dot{Q}$ between these two equations and rearranging

$$d\dot{W}_x \leqslant -\dot{m}\left[\frac{1}{\rho}\,dp + \tfrac{1}{2}d(c^2) + gdz\right]. \qquad (2.19)$$

For a turbine expansion, noting $\dot{W}_x = \dot{W}_t > 0$, integrate eqn. (2.19) from the initial state 1 to the final state 2,

$$\dot{W}_x \leqslant \dot{m} \left[\int_2^1 \frac{1}{\rho} \, dp + \tfrac{1}{2}(c_1{}^2 - c_2{}^2) + g(z_1 - z_2) \right]. \qquad (2.20)$$

For a reversible adiabatic process, $T\,ds = 0 = dh - dp/\rho$. The incremental maximum work output is then

$$d\dot{W}_{x_{\text{max}}} = -\dot{m}[dh + \tfrac{1}{2}d(c^2) + g\,dz]$$

and hence the overall maximum work output between initial state 1 and final state 2 is

$$\dot{W}_{x_{\text{max}}} = \dot{m} \int_2^1 [dh + \tfrac{1}{2}d(c^2) + g\,dz]$$

$$= \dot{m} \left[(h_{01} - h_{02s}) + g(z_1 - z_2) \right] \qquad (2.20a)$$

where the subscript s in eqn. (2.20a) denotes that the change of state between 1 and 2 is isentropic.

For an incompressible fluid, in the absence of friction, the maximum work output from the turbine (ignoring frictional losses) is

$$\dot{W}_{x_{\text{max}}} = \dot{m}g[H_1 - H_2], \qquad (2.20b)$$

where $gH = p/\rho + \tfrac{1}{2}c^2 + gz$.

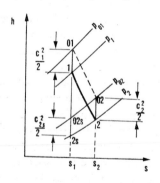

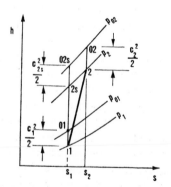

(a) Turbine expansion process   (b) Compression process

FIG. 2.5. Enthalpy–entropy diagrams for turbines and compressors.

Steam and gas turbines

Figure 2.5a shows an enthalpy–entropy or Mollier diagram on which is plotted lines of constant pressure. The process described by line 1–2 represents the expansion through an adiabatic turbine from pressure p_1 to a low pressure p_2. The ideal or reversible adiabatic expansion is represented by the line 1–2s. The fluid velocities at entry to and at exit from a turbine may be quite high and the corresponding kinetic energies may be significant. On the other hand, for a compressible fluid the potential energy terms are usually negligible. Hence the *actual* turbine rotor *specific work*

$$\Delta W_x = \dot{W}_x/\dot{m} = h_{01} - h_{02} = (h_1 - h_2) + \tfrac{1}{2}(c_1^2 - c_2^2)$$

Similarly, the *ideal* turbine rotor specific work between the same two pressures is

$$\Delta W_{x\max} = \dot{W}_{x\max}/\dot{m} = h_{01} - h_{02s} = (h_1 - h_{2s}) + \tfrac{1}{2}(c_1^2 - c_{2s}^2).$$

In Fig. 2.5a the actual turbine work/unit mass of fluid is the stagnation enthalpy change between state points 01 and 02 which lie on the stagnation pressure lines p_{01} and p_{02} respectively. The ideal turbine work per unit mass of fluid is the stagnation enthalpy change during the *isentropic process* between the same two pressures. The kinetic energy of the fluid at the end of the ideal process $\tfrac{1}{2}c_{2s}^2$ is not, however, the same as that at the end of the actual process $\tfrac{1}{2}c_2^2$. This may be adduced as follows. Taking for simplicity a perfect gas, then $h = C_pT$ and $p/\rho = RT$. Consider the constant pressure line p_2 (Fig. 2.5a); as $T_2 > T_{2s}$ then $\rho_{2s} > \rho_2$. From continuity $\dot{m}/A = \rho c$ and since we are dealing with same area, $c_2 > c_{2s}$, and the kinetic energy terms are not equal. difference in practice is usually negligible and often ignored.

There are several ways of expressing efficiency, the choice of definition depending largely upon whether the *exit kinetic energy* is usefully employed or is wasted. An example where the exhaust kinetic energy is not wasted is from the last stage of an aircraft gas turbine where it contributes to the jet propulsive thrust. Likewise, the exit kinetic energy from one stage of a multistage turbine where it is used in the next stage, provides another example. For these two cases the turbine and stage adiabatic efficiency η, is the *total-to-total efficiency* and is defined as

$$\eta_{tt} = \Delta W_x / \Delta W_{x_{max}} = (h_{01} - h_{02})/(h_{01} - h_{02s}). \qquad (2.21)$$

If the difference between the inlet and outlet kinetic energies is small, i.e. $\frac{1}{2}c_1^2 \doteqdot \frac{1}{2}c_2^2$, then

$$\eta_{tt} = (h_1 - h_2)/(h_1 - h_{2s}) \qquad (2.21a)$$

When the exhaust kinetic energy is not usefully employed and entirely wasted, the relevant adiabatic efficiency is the *total-to-static efficiency* η_{ts}. In this case the ideal turbine work is that obtained between state points 01 and 2s. Thus

$$\eta_{ts} = (h_{01} - h_{02})/(h_{01} - h_{02s} + \tfrac{1}{2}c_{2s}^2)$$
$$= (h_{01} - h_{02})/(h_{01} - h_{2s}). \qquad (2.22)$$

If the difference between inlet and outlet kinetic energies is small, eqn. (2.22) becomes

$$\eta_{ts} = (h_1 - h_2)/(h_1 - h_{2s} + \tfrac{1}{2}c_1^2). \qquad (2.22a)$$

A situation where the outlet kinetic energy is wasted is a turbine exhausting directly to the surroundings rather than through a diffuser. For example, auxiliary turbines used in rockets often do not have exhaust diffusers because the disadvantages of increased mass and space utilisation are greater than the extra propellant required as a result of reduced turbine efficiency.

Hydraulic turbines

When the working fluid is a liquid, the turbine hydraulic efficiency η_h, is defined as the work supplied by the rotor in unit time divided by the hydrodynamic energy difference of the fluid per unit time, i.e.

$$\eta_h = \frac{\Delta W_x}{\Delta W_{x_{max}}} = \frac{\Delta W_x}{g(H_1 - H_2)}. \qquad (2.23)$$

Efficiency of compressors and pumps

The *adiabatic efficiency* η_c of a compressor or the *hydraulic efficiency* of a pump η_h is broadly defined as,

$$\eta_c \ (\text{or} \ \eta_h) = \frac{\text{useful (hydrodynamic) energy input to fluid in unit time}}{\text{power input to rotor}}.$$

The power input to the rotor (or impeller) is always less than the power supplied at the coupling because of external energy losses in the bearings and glands, etc. Thus, the overall efficiency of the compressor or pump is

$$\eta_0 = \frac{\text{useful (hydrodynamic) energy input to fluid in unit time}}{\text{power input to coupling of shaft}}.$$

Hence the mechanical efficiency is

$$\eta_m = \eta_0/\eta_c \ (\text{or} \ \eta_0/\eta_h).$$

In eqn. (2.19), for a compressor or pump process, replace $-\mathrm{d}\dot{W}_x$ with $\mathrm{d}\dot{W}_c$ and rearrange the inequality to give the incremental work input

$$\mathrm{d}\dot{W}_c \geqq \dot{m}\left[\frac{1}{\rho}\,\mathrm{d}p + \tfrac{1}{2}\mathrm{d}(c^2) + g\mathrm{d}z\right]. \tag{2.24}$$

The student should carefully check the fact that the R.H.S. of this inequality is *positive*, working from eqn. (2.19).

For a complete adiabatic compression process going from state 1 to state 2, the overall work input rate is

$$\dot{W}_c \geqq \dot{m}\left[\int_1^2 \frac{\mathrm{d}p}{\rho} + \tfrac{1}{2}(c_2{}^2 - c_1{}^2) + g(z_2 - z_1)\right]. \tag{2.25}$$

For the corresponding *reversible* adiabatic compression process, noting that $T\mathrm{d}s = 0 = \mathrm{d}h - \mathrm{d}p/\rho$, the minimum work input rate is

$$\dot{W}_{c\min} = \dot{m}\int_1^{2s} [\mathrm{d}h + \tfrac{1}{2}\mathrm{d}c^2 + g\mathrm{d}z]$$

$$= \dot{m}[(h_{02s} - h_{01}) + g(z_2 - z_1)]. \tag{2.26}$$

From the steady flow energy equation, for an adiabatic process in a compressor

$$\dot{W}_c = \dot{m}(h_{02} - h_{01}). \tag{2.27}$$

Figure 2.5b shows a Mollier diagram on which the actual compression process is represented by the state change 1–2 and the corresponding ideal process by 1–2s. For an adiabatic compressor the only meaningful efficiency is the total-to-total efficiency which is

$$\eta_c = \frac{\text{minimum adiabatic work input per unit time}}{\text{actual adiabatic work input to rotor per unit time}}$$

$$= \frac{h_{02s} - h_{01}}{h_{02} - h_{01}}. \tag{2.28}$$

If the difference between inlet and outlet kinetic energies is small, $\frac{1}{2}c_1^2 \doteq \frac{1}{2}c_2^2$ and

$$\eta_c = \frac{h_{2s} - h_1}{h_2 - h_1}. \tag{2.28a}$$

For *incompressible* flow, eqn. (2.25) gives

$$\Delta W_p = \dot{W}_p/m \geq [(p_2 - p_1)/\rho + \tfrac{1}{2}(c_2^2 - c_1^2) + g(z_2 - z_1)]$$
$$\geq g[H_2 - H_1].$$

For the ideal case with no fluid friction

$$\Delta W_{p_{\min}} = g[H_2 - H_1]. \tag{2.29}$$

For a pump the hydraulic efficiency is defined as

$$\eta_h = \frac{\Delta W_{p_{\min}}}{\Delta W_p} = \frac{g[H_2 - H_1]}{\Delta W_p}. \tag{2.30}$$

Small stage or polytropic efficiency

The adiabatic efficiency described in the preceding section, although fundamentally valid, can be misleading if used for comparing the efficiencies of turbomachines of differing pressure ratios. Now any turbomachine may be regarded as being composed of a large number of very small stages irrespective of the actual number of stages in the machine. If each small stage has the same efficiency, then the adiabatic efficiency of the whole machine will be different from the small stage efficiency, the difference depending upon the pressure ratio of the machine. This rather surprising result is a manifestation of a simple thermodynamic effect concealed in the expression for adiabatic efficiency and is made apparent in the following argument.

Compression process

Figure 2.6 shows an enthalpy–entropy diagram on which adiabatic compression between pressures p_1 and p_2 is represented by the change of state between points 1 and 2. The corresponding reversible process is represented by the isentropic line 1 to 2s. It is assumed that the compression process may be divided up into a large number of small stages of equal efficiency η_p. For each small stage the actual work input is $\Delta \dot{W}$ and the corresponding ideal work in the isentropic process is $\Delta \dot{W}_{\min}$. With the notation of Fig. 2.6,

$$\eta_p = \frac{\Delta \dot{W}_{\min}}{\Delta \dot{W}} = \frac{h_{xs} - h_1}{h_x - h_1} = \frac{h_{ys} - h_x}{h_y - h_x} = \dots$$

Since each small stage has the same efficiency, then $\eta_p = (\Sigma \Delta \dot{W}_{\min})/(\Sigma \Delta \dot{W})$ is also true.

From the relation $T\mathrm{d}s = \mathrm{d}h - v\mathrm{d}p$, for a constant pressure process $(\partial h/\partial s)_p = T$. This means that the higher the fluid temperature *greater* is the slope of the constant pressure lines on the Mollier diagram. For a gas where h is a function of T, constant pressure lines diverge and the slope of the line p_2 is greater than the slope of line p_1 at the same value of entropy. At equal values of T, constant pressure lines are of equal slope as indicated in Fig. 2.6. For the special case of a *perfect gas* (where C_p is constant), $C_p(\mathrm{d}T/\mathrm{d}s) = T$ for a constant pressure process. Integrating this expression results in the equation for a constant pressure line, $s = C_p \log T +$ constant.

Returning now to the more general case, since

$$(\Sigma \Delta \dot{W})/\dot{m} = \{(h_x - h_1) + (h_y - h_x) + \dots\} = (h_2 - h_1),$$

then

$$\eta_p = [(h_{xs} - h_1) + (h_{ys} - h_x) + \dots]/(h_2 - h_1).$$

The adiabatic efficiency of the *whole* compression process is

$$\eta_c = (h_{2s} - h_1)/(h_2 - h_1).$$

Because of the divergence of the constant pressure lines

$$\{(h_{xs} - h_1) + (h_{ys} - h_x) + \ldots\} > (h_{2s} - h_1\},$$

i.e.

$$\Sigma\Delta\dot{W}_{min} > \dot{W}_{min}.$$

Therefore,

$$\eta_p > \eta_c.$$

Thus, for a compression process the adiabatic efficiency of the machine is *less* than the small stage efficiency, the difference being dependent upon the divergence of the constant pressure lines. Although the fore-

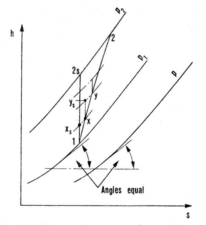

FIG. 2.6. Compression process by small stages.

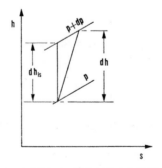

FIG. 2.7. Incremental change of state in a compression process.

going discussion has been in terms of static states it can be regarded as applying to stagnation states if the inlet and outlet kinetic energies from each stage are equal.

Small stage efficiency for a perfect gas

An explicit relation can be readily derived for a perfect gas (C_p is constant) between small stage efficiency, the overall adiabatic efficiency and pressure ratio. The analysis is for the limiting case of an infinitesimal compressor stage in which the incremental change in pressure is dp as indicated in Fig. 2.7. For the actual process the incremental enthalpy rise is dh and the corresponding ideal enthalpy rise is dh_{is}.

The polytropic efficiency for the small stage is

$$\eta_p = \frac{dh_{is}}{dh} = \frac{v\,dp}{C_p\,dT},\qquad(2.31)$$

since for an isentropic process $T\,ds = 0 = dh_{is} - v\,dp$.

Substituting $v = RT/p$ in eqn. (2.31), then

$$\eta_p = \frac{R}{C_p}\frac{T}{p}\frac{dp}{dT}$$

and hence

$$\frac{dT}{T} = \frac{(\gamma - 1)}{\gamma\eta_p}\frac{dp}{p}\qquad(2.32)$$

as
$$C_p = \gamma R/(\gamma - 1).$$

Integrating eqn. (2.32) across the whole compressor and taking equal efficiency for each infinitesimal stage gives,

$$\frac{T_2}{T_1} = \left(\frac{p_2}{p_1}\right)^{(\gamma - 1)/\eta_p\gamma}.\qquad(2.33)$$

Now the adiabatic efficiency for the whole compression process is

$$\eta_c = (T_{2s} - T_1)/(T_2 - T_1)\qquad(2.34)$$

if it is assumed that the velocities at inlet and outlet are equal.

For the *ideal* compression process put $\eta_p = 1$ in eqn. (2.32) and so obtain

$$\frac{T_{2s}}{T_1} = \left(\frac{p_2}{p_1}\right)^{(\gamma-1)/\gamma} \tag{2.35}$$

which is also obtainable from $pv^\gamma = \text{constant}$ and $pv = RT$. Substituting eqns. (2.33) and (2.35) into eqn. (2.34) results in the expression

$$\eta_c = \left[\left(\frac{p_2}{p_1}\right)^{(\gamma-1)/\gamma} - 1\right]\bigg/\left[\left(\frac{p_2}{p_1}\right)^{(\gamma-1)/\eta_p\gamma} - 1\right]. \tag{2.36}$$

Values of "overall" adiabatic efficiency have been calculated using eqn. (2.36) for a range of pressure ratio and different values of η_p, and are plotted in Fig. 2.8. This figure amplifies the observation made

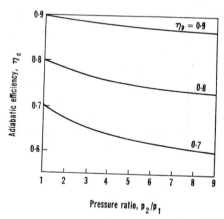

Fig. 2.8. Relationship between adiabatic (overall) efficiency, pressure ratio and small stage (polytropic) efficiency for a compressor ($\gamma = 1.4$).

earlier that the adiabatic efficiency of a finite compression process is *less* than the efficiency of the small stages. Comparison of the adiabatic efficiency of two machines of different pressure ratios is not a valid procedure since, for equal polytropic efficiency, the compressor with the highest pressure ratio is penalised by the hidden thermodynamic effect.

The term *polytropic* used above arises in the context of a reversible compressor compressing a gas from the same initial state to the same final state as the irreversible adiabatic compressor but obeying the relation $pv^n = $ constant. The index n is called the *polytropic index*. Since an increase in entropy occurs for the change of state in both compressors, for the reversible compressor this is only possible if there is a reversible heat transfer $dQ_R = Tds$. Proceeding farther, it follows that the value of the index n must always exceed that of γ. This is clear from the following argument. For the polytropic process,

$$dQ_R = du + pdv.$$

$$= \frac{C_v}{R} d(pv) + pdv.$$

Using $pv^n = $ constant and $C_v = R/(\gamma-1)$, after some manipulation the expression $dQ_R = (\gamma-n)/(\gamma-1)pdv$ is derived. For a compression process $dv < 0$ and $dQ_R > 0$ then $n > \gamma$. For an expansion process $dv > 0$, $dQ_R < 0$ and again $n > \gamma$.

Turbine polytropic efficiency

A similar analysis to the compression process can be applied to a perfect gas expanding through an adiabatic turbine. For the turbine the appropriate expressions for an expansion, from a state 1 to a state 2, are

$$\frac{T_2}{T_1} = \left(\frac{p_2}{p_1}\right)^{\eta_p(\gamma-1)/\gamma}, \tag{2.37}$$

$$\eta_t = \left[1 - \left(\frac{p_2}{p_1}\right)^{\eta_p(\gamma-1)/\gamma}\right] \Big/ \left[1 - \left(\frac{p_2}{p_1}\right)^{(\gamma-1)/\gamma}\right]. \tag{2.38}$$

The derivation of these expressions is left as an exercise for the student. "Overall" adiabatic efficiencies have been calculated for a range of pressure ratio and different polytropic efficiencies and are shown in Fig. 2.9. The most notable feature of these results is that, in contrast with a compression process, for an expansion, adiabatic efficiency *exceeds* small stage efficiency.

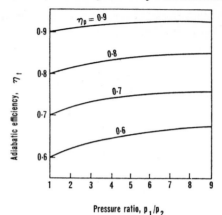

FIG. 2.9. Turbine adiabatic efficiency against pressure ratio for various polytropic efficiencies ($\gamma = 1\cdot4$).

Reheat factor

The foregoing relations obviously cannot be applied to steam turbines as vapours do not obey the gas laws. It is customary in steam turbine practice to use a *reheat factor* R_H as a measure of the inefficiency of the complete expansion. Referring to Fig. 2.10, the expansion process through an adiabatic turbine from state 1 to state 2 is shown on a Mollier diagram, split into a number of small stages. The reheat factor is defined as

$$R_H = [(h_1 - h_{xs}) + (h_x - h_{ys}) + \ldots]/(h_1 - h_{2s}) = (\Sigma \Delta h_{is})/(h_1 - h_{2s}).$$

Due to the gradual divergence of the constant pressure lines on the *h–s* diagram, R_H is greater than unity. The actual value of R_H for an infinite number of stages depends upon the position of the expansion line upon the Mollier diagram; it is usually between $1\cdot03$ to $1\cdot08$ in normal steam turbine practice.

Now since the adiabatic efficiency of the turbine is

$$\eta_t = \frac{h_1 - h_2}{h_1 - h_{2s}} = \frac{h_1 - h_2}{\Sigma \Delta h_{is}} \cdot \frac{\Sigma \Delta h_{is}}{h_1 - h_{2s}}$$

then

$$\eta_t = \eta_p R_H \qquad (2.39)$$

which establishes the connection between polytropic efficiency, reheat factor and turbine adiabatic efficiency.

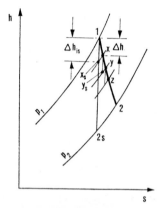

FIG. 2.10. Mollier diagram showing expansion process through an adiabatic turbine.

Nozzle efficiency

In a large number of turbomachinery components the flow process can be regarded as a purely nozzle flow in which the fluid receives an acceleration as a result of a drop in pressure. Such a nozzle flow occurs at entry to all turbomachines and in the stationary blade rows in turbines. In axial machines the expansion at entry is assisted by a row of stationary blades (called *guide vanes* in compressors and *nozzles* in turbines) which direct the fluid on to the rotor with a large swirl angle. Centrifugal compressors and pumps, on the other hand, often have no such provision for flow guidance but there is still a velocity increase obtained from a contraction in entry flow area. Figure 2.11a shows the process represented on a Mollier diagram, the expansion proceeding from state 1 to state 2. The process can be regarded as being adiabatic and, as there is no shaft work, the steady flow energy equation gives

$$h_{01} = h_{02} \quad \text{or} \quad h_1 - h_2 = \tfrac{1}{2}(c_2^2 - c_1^2). \tag{2.40}$$

For the equivalent reversible adiabatic process

$$h_1 - h_{2s} = \tfrac{1}{2}(c_{2s}^2 - c_1^2).$$

The nozzle efficiency can be defined as

$$\eta_N = (h_1 - h_2)/(h_1 - h_{2s}) = (c_2^2 - c_1^2)/(c_{2s}^2 - c_1^2). \qquad (2.41)$$

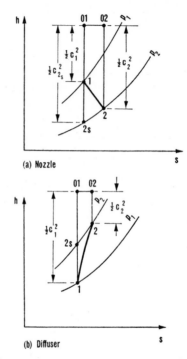

(a) Nozzle

(b) Diffuser

Fig. 2.11. Mollier diagrams for nozzle and diffuser processes.

Now for an isentropic change, $Tds = dh - vdp = 0$. If the fluid is, or can be treated as, incompressible, the change in state from 1 to $2s$ can be written

$$(h_1 - h_{2s}) = (p_1 - p_2) \div \rho. \qquad (2.42)$$

Subtracting eqn. (2.40) from eqn. (2.42)

$$h_2 - h_{2s} = (p_1 - p_2) \div \rho - \tfrac{1}{2}(c_2^2 - c_1^2)$$
$$= (p_{01} - p_{02}) \div \rho. \qquad (2.43)$$

Thus, substituting eqns. (2.42) and (2.43) into eqn. (2.41) the nozzle efficiency for incompressible flow is

$$\eta_N = 1 - \frac{p_{01} - p_{02}}{p_1 - p_2}. \tag{2.44}$$

Diffuser efficiency

Figure 2.11b shows a diffusion process represented on a Mollier diagram by the change of state from points 1 to 2. For steady adiabatic flow in stationary passages, $h_{01} = h_{02}$ and, therefore,

$$h_2 - h_1 = \tfrac{1}{2}(c_1^2 - c_2^2). \tag{2.45}$$

For the equivalent reversible adiabatic process from 1 to 2s

$$h_{2s} - h_1 = \tfrac{1}{2}(c_1^2 - c_{2s}^2). \tag{2.46}$$

The diffuser efficiency η_D, is defined in an analogous manner to nozzle efficiency as,

$$\eta_D = (h_{2s} - h_1)/(h_2 - h_1) = (c_1^2 - c_{2s}^2)/(c_1^2 - c_2^2). \tag{2.47}$$

For flow which is incompressible (or can be regarded as nearly incompressible)

$$h_{2s} - h_1 = (p_2 - p_1) \div \rho,$$

hence,

$$\eta_D = \frac{2(p_2 - p_1)}{\rho(c_1^2 - c_2^2)}. \tag{2.48}$$

Equation (2.48) can be expressed in terms of pressure changes alone since,

$$(h_2 - h_{2s}) = (h_2 - h_1) - (h_{2s} - h_1)$$
$$= \tfrac{1}{2}(c_1^2 - c_2^2) - (p_2 - p_1) \div \rho$$
$$= (p_{01} - p_{02}) \div \rho,$$

then

$$\eta_D = \frac{(h_{2s} - h_1)}{(h_{2s} - h_1) - (h_{2s} - h_2)} = \frac{1}{1 - (h_{2s} - h_2)/(h_{2s} - h_1)}$$

$$= \frac{1}{1 + (p_{01} - p_{02})/(p_2 - p_1)}. \tag{2.49}$$

The diffusion or deceleration of fluid is an essential feature of some part of most turbomachines and has as its aim the efficient conversion of kinetic energy to pressure energy. This aim is difficult to accomplish and is rightly regarded as one of the central problems of turbo-machinery design. The difficulty stems from the fact that the boundary layer is prone to separation if the rate of diffusion is too rapid, and large losses in stagnation pressure are then inevitable. On the other hand, if the rate of diffusion is very low the fluid is exposed to an excessive length of wall and friction losses become predominant. Clearly, there must be an optimum rate of diffusion for which the two effects are minimised. Equation (2.49) expresses the fact that when diffuser efficiency η_D is a maximum, the total pressure loss is a *minimum* for a given rise in static pressure. A general discussion of the optimum design of straight-walled diffusers with various geometries is given by Kline et al.[5] They correlated data from many sources and showed, for the diffuser geometries given in Fig. 2.12, that the optimum diffuser efficiency occurs when the included angle 2θ is about 7 deg. However, several other important optimum problems are frequently encountered and these have been discussed in some detail by Kline and his co-authors.

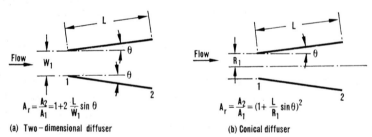

$$A_r = \frac{A_2}{A_1} = 1 + 2\frac{L}{W_1}\sin\theta$$

(a) Two–dimensional diffuser

$$A_r = \frac{A_2}{A_1} = \left(1 + \frac{L}{R_1}\sin\theta\right)^2$$

(b) Conical diffuser

Fig. 2.12. Subsonic diffuser geometries and area ratios.

One such optimum problem is the requirement of maximum pressure recovery for a given diffuser length in the flow direction regardless of the area ratio $A_r = A_2/A_1$. This may seem surprising but, in general, this optimum condition produces a different diffuser geometry from that needed for optimum efficiency. This can be demonstrated by means of the following considerations.

For an incompressible flow through a diffuser the energy equation can be written as

$$\frac{p_1}{\rho} + \tfrac{1}{2} c_1^2 = \frac{p_2}{\rho} + \tfrac{1}{2} c_2^2 + \frac{\Delta p_0}{\rho} \qquad (2.50)$$

where the loss in total pressure $\Delta p_0 = p_{01} - p_{02}$. A pressure rise coefficient can be defined as $C_p = (p_2 - p_1)/q_1$ where $q_1 = \tfrac{1}{2}\rho c_1^2$. From eqn. (2.50) it is easy to show that the *ideal* pressure rise coefficient is

$$C_{pi} = 1 - (c_2/c_1)^2$$

by setting Δp_0 to zero. Thus, eqn. (2.50) can be written as

$$C_p = C_{pi} - \Delta p_0/q_1. \qquad (2.51)$$

The diffuser efficiency, using the definition in eqn. (2.48), is

$$\eta_D = C_p/C_{pi} \qquad (2.48a)$$

hence

$$\ln \eta_D = \ln C_p - \ln C_{pi}.$$

Differentiating this equation with respect to θ and setting the result to zero, the condition for maximum efficiency is obtained, i.e.

$$\frac{1}{C_p} \frac{\partial C_p}{\partial \theta} = \frac{1}{C_{pi}} \frac{\partial C_{pi}}{\partial \theta}. \qquad (2.52)$$

Thus, at maximum efficiency the fractional rate of increase of C_p with angle is equal to the fractional rate of increase of C_{pi} with angle. As C_p at this point is positive and, by definition, both C_{pi} and $\partial C_{pi}/\partial \theta$ are also positive, eqn. (2.52) shows that $\partial C_p/\partial \theta > 0$ at the maximum efficiency point. Clearly, C_p *cannot* be at its maximum when η_d is at its peak value. C_p continues to increase until $\partial C_p/\partial \theta = 0$.

Now differentiating eqn. (2.51) with respect to θ and setting the L.H.S. to zero, the condition for maximum C_p is obtained, viz.

$$\frac{\partial C_{pi}}{\partial \theta} = \frac{\partial}{\partial \theta} (\Delta p_0/q_1).$$

Thus, as the diffuser angle is increased beyond the divergence which gave maximum efficiency, the actual pressure rise will continue to rise until the additional losses in total pressure balance the theoretical gain in pressure recovery produced by the increased area ratio.

At bigger included angles large transitory stall occurs in which the separation varies in position, size and intensity with time. A companion paper by Kline[6] is of especial interest to those students wishing to acquire a simple physical understanding of what occurs in stalling flows, particularly passage flows. In Chapter 3 of this book the special problems of diffusion and stall within axial compressor blade rows is treated using the correlation methods of Lieblein and Howell.

Figure 2.13 shows a typical design chart for straight-walled diffusers on which is shown the lines of optimum efficiency γ–γ and optimum recovery at constant L/W_1. The line a–a denotes approximately the limiting angle of divergence 2θ for any value of L/W_1 above which the large transitory stall phenomenon appears.

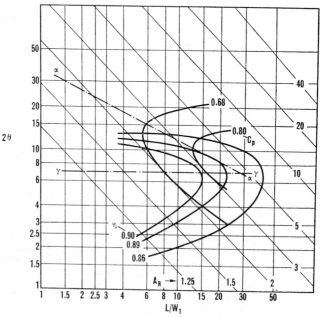

Fig. 2.13. Typical design curve, straight-walled diffusers (Kline *et al.*[5]).
(Courtesy Am. Soc. Mech. Engrs.).

REFERENCES

1. GIBBINGS, J. C., *Thermomechanics*; *the Governing Equations*. Pergamon Press Oxford (1970).
2. MONTGOMERY, S. R., *The Second Law of Thermodynamics*. Pergamon Press, Oxford (1966).
3. KEENAN, J. H., *Thermodynamics*. Wiley, New York (1941).
4. SPALDING, D. B. and COLE, E. H., *Engineering Thermodynamics*. Arnold, London (1973).
5. KLINE, S. J., ABBOTT, D. E. and FOX, R. W., Optimum design of straight-walled diffusers. *Trans. Am. Soc. Mech. Engrs.* Series D, **81**, (1959).
6. KLINE, S. J., On the nature of stall. *Trans. Am. Soc. Mech. Engrs.* Series D, **81**, (1959).

PROBLEMS

1. For the adiabatic expansion of a perfect gas through a turbine, show that the overall efficiency η_t and small stage efficiency η_r are related by

$$\eta_t = (1 - \varepsilon^{\eta_p})/(1 - \epsilon),$$

where $\epsilon = r^{(1-\gamma)/\gamma}$, and r is the expansion pressure ratio, γ is the ratio of specific heats.

An axial flow turbine has a small stage efficiency of 86%, an overall pressure ratio of 4·5 to 1 and a mean value of γ equal to 1·333. Calculate the overall turbine efficiency.

2. Air is expanded in a multi-stage axial flow turbine, the pressure drop across each stage being very small. Assuming that air behaves as a perfect gas with ratio of specific heats γ, derive pressure–temperature relationships for the following processes:

(i) reversible adiabatic expansion;
(ii) irreversible adiabatic expansion, with small stage efficiency η_p;
(iii) reversible expansion in which the heat loss in each stage is a constant fraction k of the enthalpy drop in that stage;
(iv) reversible expansion in which the heat loss is proportional to the absolute temperature T.

Sketch the first three processes on a T,s diagram.

If the entry temperature is 1100 K, and the pressure ratio across the turbine is 6 to 1, calculate the exhaust temperatures in each of these three cases. Assume that γ is 1·333, that $\eta_p = 0·85$, and that $k = 0·1$.

3. A multi-stage high-pressure steam turbine is supplied with steam at a stagnation pressure of 7 MPa abs. and a stagnation temperature of 500°C. The corresponding specific enthalpy is 3410 kJ/kg. The steam exhausts from the turbine at a stagnation pressure of 0·7 MPa abs., the steam having been in a superheated condition throughout the expansion. It can be assumed that the steam behaves like a perfect gas over the range of the expansion and that $\gamma = 1·3$. Given that the turbine flow process has a small-stage efficiency of 0·82, determine

(i) the temperature and specific volume at the end of the expansion;
(ii) the reheat factor.

The specific volume of superheated steam is represented by $pv = 0.231(h - 1943)$, where p is in kPa, v is in m^3/kg and h is in kJ/kg.

4. A 20 MW back-pressure turbine receives steam at 4 MPa abs. and 300°C, exhausting from the last stage at 0.35 MPa. The stage efficiency is 0.85, reheat factor 1.04 and external losses 2% of the insentropic enthalpy drop. Determine the rate of steam flow.

At the exit from the first stage nozzles the steam velocity is 244 m/s, specific volume 68.6 *l*/kg, mean diameter 762 mm and steam exit angle 76 deg measured from the axial direction. Determine the nozzle exit height of this stage.

5. Steam is supplied to the first stage of a five stage pressure-compounded steam turbine at a stagnation pressure of 1.5 MPa abs. and a stagnation temperature of 350°C. The steam leaves the last stage at a stagnation pressure of 7.0 kPa abs. with a corresponding dryness fraction of 0.95. By using a Mollier chart for steam and assuming that the stagnation state point locus is a straight line joining the initial and final states, determine

  (i) the stagnation conditions between each stage assuming that each stage does the same amount of work;
 (ii) the total-to-total efficiency of each stage;
(iii) the overall total-to-total efficiency and total-to-static efficiency assuming the steam enters the condenser with a velocity of 200 m/s;
(iv) the reheat factor based upon stagnation conditions.

CHAPTER 3

Two-dimensional Cascades

Let us first understand the facts and then we may seek the causes. (ARISTOTLE.)

THE operation of any turbomachine is directly dependent upon changes in the working fluid's angular momentum as it crosses individual blade rows. A deeper insight of turbomachinery mechanics may be gained from consideration of the flow changes and forces exerted within these individual blade rows. In this chapter the flow past twô-dimensional blade cascades is examined.

A cascade tunnel is shown in Fig. 3.1, the cascade itself comprising a number of identical blades, equally spaced and parallel to one another.

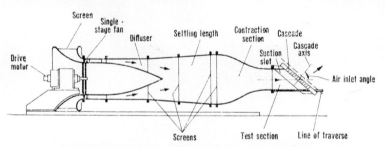

FIG. 3.1. Layout of a conventional low-speed cascade tunnel (Lieblen,[17] Ch. 6). (Courtesy of NASA.)

To obtain truly two-dimensional flow would require a cascade of infinite extent. Of necessity cascades must be limited in size, and careful design is needed to ensure that at least the central regions (where flow measurements are made) operate with approximately two-dimensional flow.

53

For axial flow machines of high hub-tip ratio, radial velocities are negligible and, to a close approximation, the flow may be described as two-dimensional. The flow in a cascade is then a reasonable model of the flow in the machine. With lower hub-tip radius ratios, the blades of a turbomachine will normally have an appreciable amount of twist along their length, the amount depending upon the sort of "vortex design" chosen (see Chapter 6). However, data obtained from two-dimensional cascades can still be of value to a designer requiring the performance at discrete blade sections of such blade rows.

CASCADE NOMENCLATURE

A cascade blade profile can be conceived as a curved *camber line* upon which a *profile thickness distribution* is symmetrically superimposed. Referring to Fig. 3.2 the camber line $y(x)$ and profile thickness $t(x)$ are shown as functions of the distance x along the *blade chord l*. In British

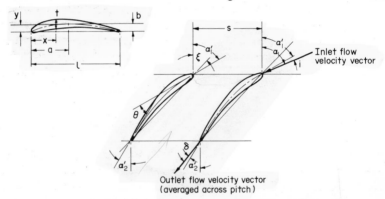

Fig. 3.2. Compressor cascade and blade notation.

practice the shape of the camber line is usually either a circular arc or a parabolic arc defined by the maximum camber b located at distance a from the leading edge of the blade. The profile thickness distribution may be that of a standard aerofoil section but, more usually, is one of the sections specifically developed by the various research establishments for compressor or turbine applications. Blade camber and thickness distributions are generally presented as tables of y/l and t/l against

x/l. Some examples of these tables are quoted by Horlock.[4, 25] Summarising, the useful parameters for describing a cascade blade are: camber line shape, b/l, a/l, type of thickness distribution and maximum thickness to chord ratio, t_{max}/l.

With the blades arranged in cascade, two important additional geometric variables which define the cascade are the *space-chord ratio* s/l and the *stagger angle* ξ, which is the angle between the chord line and a reference direction *perpendicular to the cascade front*. Throughout the remainder of this book, all fluid and blade angles are referred to this perpendicular so as to avoid the needless complication arising from the use of other reference directions. However, custom dies hard; in steam turbine practice, blade and flow angles are conventionally measured from the *tangential* direction (i.e. parallel to the cascade front). Despite this, it is better to avoid ambiguity of meaning by adopting the single reference direction already given.

The blades angles at entry to and at exit from the cascade are denoted by α_1' and α_2' respectively. A most useful blade parameter is the *camber angle* θ which is the change in angle of the camber line between the leading and trailing edges and equals $\alpha_1' - \alpha_2'$ in the notation of Fig. 3.2. For circular arc camber lines the stagger angle is $\xi = \frac{1}{2}(\alpha_1' + \alpha_2')$. For parabolic arc camber lines of low camber (i.e. small b/l) as used in some compressor cascades, the inlet and outlet blade angles are

$$\alpha_1' = \xi + \tan^{-1}\frac{b/l}{(a/l)^2} \qquad \alpha_2' = \xi - \tan^{-1}\frac{b/l}{(1 - a/l)^2}$$

the equation approximating for the parabolic arc being $Y = X\{A(X - 1) + BY\}$ where $X = x/l$, $Y = y/l$. A, B are two arbitrary constants which can be solved with the conditions that at $x = a$, $y = b$ and $dy/dx = 0$. The exact general equation of a parabolic arc camber line which has been used in the design of highly cambered turbine blades is dealt with by Dunham.[26]

ANALYSIS OF CASCADE FORCES

The fluid approaches the cascade from far upstream with velocity c_1 at an angle α_1 and leaves far downstream of the cascade with velocity

c_2 at an angle α_2. In the following analysis the fluid is assumed to be incompressible and the flow to be steady. The assumption of steady flow is valid for an isolated cascade row but, in a turbomachine, relative motion between successive blade rows gives rise to unsteady flow effects. As regards the assumption of incompressible flow, the majority of cascade tests are conducted at fairly low Mach numbers (e.g. 0·3 on compressor cascades) when compressibility effects are negligible. Various techniques are available for correlating incompressible and compressible cascades; a brief review is given by Csanady.[1]

A portion of an isolated blade cascade (for a compressor) is shown in Fig. 3.3. The forces X and Y are exerted by unit depth of blade upon the fluid, exactly equal and opposite to the forces exerted by the fluid upon unit depth of blade. A control surface is drawn with end bound-

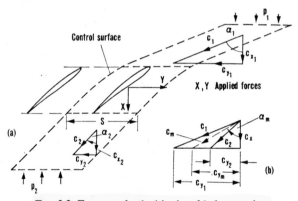

Fig. 3.3. Forces and velocities in a blade cascade.

aries far upstream and downstream of the cascade and with side boundaries coinciding with the median stream lines.

Applying the principle of continuity to a unit depth of span and noting the assumption of incompressibility, yields

$$c_1 \cos \alpha_1 = c_2 \cos \alpha_2 = c_x. \tag{3.1}$$

The momentum equation applied in the x and y directions with constant axial velocity gives,

$$X = (p_2 - p_1)s, \tag{3.2}$$

$$Y = \rho s c_x (c_{y1} - c_{y2}), \tag{3.3}$$

or

$$Y = \rho s c_x^2 (\tan \alpha_1 - \tan \alpha_2). \tag{3.3a}$$

Equations (3.1) and (3.3) are completely valid for a flow incurring total pressure losses in the cascade.

ENERGY LOSSES

A real fluid crossing the cascade experiences a loss in total pressure Δp_0 due to skin friction and related effects. Thus

$$\frac{\Delta p_0}{\rho} = \frac{p_1 - p_2}{\rho} + \tfrac{1}{2}(c_1^2 - c_2^2). \tag{3.4}$$

Noting that $c_1^2 - c_2^2 = (c_{y1}^2 + c_x^2) - (c_{y2}^2 + c_x^2) = (c_{y1} + c_{y2})(c_{y1} - c_{y2})$, substitute eqns. (3.2) and (3.3) into eqn. (3.4) to derive the relation,

$$\frac{\Delta p_0}{\rho} = \frac{1}{\rho s}(-X + Y \tan \alpha_m), \tag{3.5}$$

where

$$\tan \alpha_m = \tfrac{1}{2}(\tan \alpha_1 + \tan \alpha_2). \tag{3.6}$$

A non-dimensional form of eqn. (3.5) is often useful in presenting the results of cascade tests. Several forms of total pressure-loss coefficient can be defined of which the most popular are,

$$\zeta = \Delta p_0 / (\tfrac{1}{2}\rho c_x^2) \tag{3.7a}$$

and

$$\bar{\omega} = \Delta p_0 / (\tfrac{1}{2}\rho c_1^2). \tag{3.7b}$$

Using again the same reference parameter, a pressure rise coefficient C_p and a tangential force coefficient C_f may be defined

$$C_p = \frac{p_2 - p_1}{\tfrac{1}{2}\rho c_x^2} = \frac{X}{\tfrac{1}{2}\rho s c_x^2}, \tag{3.8}$$

$$C_f = \frac{Y}{\tfrac{1}{2}\rho s c_x^2} = 2(\tan \alpha_1 - \tan \alpha_2), \tag{3.9}$$

using eqns. (3.2) and (3.3a).

Substituting these coefficients into eqn. (3.5) to give, after some rearrangement,

$$C_p = C_f \tan \alpha_m - \zeta. \tag{3.10}$$

LIFT AND DRAG

A mean velocity c_m is defined as

$$c_m = c_x/\cos \alpha_m, \tag{3.11}$$

where α_m is itself defined by eqn. (3.6). Considering unit depth of a cascade blade, a lift force L acts in a direction perpendicular to c_m and a drag force D in a direction parallel to c_m. Figure 3.4 shows L and D as the reaction forces exerted *by* the blade *upon* the fluid.

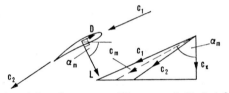

FIG. 3.4. Lift and drag forces exerted by a cascade blade (of unit span) upon the fluid.

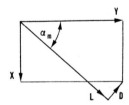

FIG. 3.5. Axial and tangential forces exerted by unit span of a blade upon the fluid.

Experimental data are often presented in terms of lift and drag when the data may be of greater use in the form of tangential force and total pressure loss. The lift and drag forces can be resolved in terms of the axial and tangential forces. Referring to Fig. 3.5,

$$L = X \sin \alpha_m + Y \cos \alpha_m, \tag{3.12}$$

$$D = Y \sin \alpha_m - X \cos \alpha_m. \tag{3.13}$$

From eqn. (3.5)

$$D = \cos \alpha_m (Y \tan \alpha_m - X) = s\Delta p_0 \cos \alpha_m. \tag{3.14}$$

Rearranging eqn. (3.14) for X and substituting into eqn. (3.12) gives,

$$
\begin{aligned}
L &= (Y \tan \alpha_m - s\Delta p_0) \sin \alpha_m + Y \cos \alpha_m \\
&= Y \sec \alpha_m - s\Delta p_0 \sin \alpha_m \\
&= \rho s c_x^2 (\tan \alpha_1 - \tan \alpha_2) \sec \alpha_m - s\Delta p_0 \sin \alpha_m, \tag{3.15}
\end{aligned}
$$

after using eqn. (3.9).

Lift and drag coefficients may be introduced as

$$C_L = \frac{L}{\frac{1}{2}\rho c_m^2 l}, \tag{3.16a}$$

$$C_D = \frac{D}{\frac{1}{2}\rho c_m^2 l}. \tag{3.16b}$$

Using eqn. (3.14) together with eqn. (3.7),

$$C_D = \frac{s\Delta p_0 \cos \alpha_m}{\frac{1}{2}\rho c_m^2 l} = \zeta \frac{s}{l} \cos^3 \alpha_m. \tag{3.17}$$

With eqn. (3.15)

$$C_L = \frac{\rho s c_x^2 (\tan \alpha_1 - \tan \alpha_2) \sec \alpha_m - s\Delta p_0 \sin \alpha_m}{\frac{1}{2}\rho c_m^2 l}$$

$$= 2 \frac{s}{l} \cos \alpha_m (\tan \alpha_1 - \tan \alpha_2) - C_D \tan \alpha_m. \tag{3.18}$$

Alternatively, employing eqns. (3.9) and (3.17),

$$C_L = \frac{s}{l} \cos \alpha_m \left(C_f - \zeta \frac{\sin 2\alpha_m}{2} \right). \tag{3.19}$$

Within the normal range of operation in a cascade, values of C_D are very much less than C_L. As α_m is unlikely to exceed 60 deg, the quantity $C_D \tan \alpha_m$ in eqn. (3.18) can be dropped, resulting in the approximation,

$$\frac{L}{D} = \frac{C_L}{C_D} \doteq \frac{2 \sec^2 \alpha_m}{\zeta} (\tan \alpha_1 - \tan \alpha_2) = \frac{C_f}{\zeta} \sec^2 \alpha_m. \tag{3.20}$$

CIRCULATION AND LIFT

The lift of a single isolated aerofoil for the ideal case when $D = 0$ is given by the Kutta–Joukowski theorem

$$L = \rho \Gamma c, \tag{3.21}$$

where c is the relative velocity between the aerofoil and the fluid at infinity and Γ is the circulation about the aerofoil. This theorem is of fundamental importance in the development of the theory of aerofoils (for further information see Glauert[2]).

In the absence of total pressure losses, the lift force per unit span of a blade *in cascade*, using eqn. (3.15), is

$$L = \rho s c_x^2 (\tan a_1 - \tan a_2) \sec a_m$$

$$= \rho s c_m (c_{y1} - c_{y2}). \tag{3.22}$$

Now the *circulation* is the contour integral of velocity around a closed curve. For the cascade blade the circulation is

$$\Gamma = s(c_{y1} - c_{y2}). \tag{3.23}$$

Combining eqns. (3.22) and (3.23),

$$L = \rho \Gamma c_m. \tag{3.24}$$

As the spacing between the cascade blades is increased without limit (i.e. $s \to \infty$), the inlet and outlet velocities to the cascade, c_1 and c_2, become equal in magnitude and direction. Thus $c_1 = c_2 = c$ and eqn. (3.24) becomes identical with the Kutta–Joukowski theorem obtained for an isolated aerofoil.

EFFICIENCY OF A COMPRESSOR CASCADE

The efficiency η_D of a compressor blade cascade can be defined in the same way as diffuser efficiency; this is the ratio of the actual static pressure rise in the cascade to the maximum possible theoretical pressure rise (i.e. with $\Delta p_0 = 0$). Thus,

$$\eta_D = \frac{p_2 - p_1}{\frac{1}{2}\rho(c_1^2 - c_2^2)}$$

$$= 1 - \frac{\Delta p_0}{\rho c_x^2 \tan \alpha_m (\tan \alpha_1 - \tan \alpha_2)}.$$

Inserting eqns. (3.7) and (3.9) into the above equation,

$$\eta_D = 1 - \frac{\zeta}{C_f \tan \alpha_m}. \tag{3.25}$$

Equation (3.20) can be written as $\zeta/C_f \doteq (\sec^2 \alpha_m) C_D / C_L$ which when substituted into eqn. (3.25) gives

$$\eta_D = 1 - \frac{2C_D}{C_L \sin 2\alpha_m}. \tag{3.26}$$

Assuming a constant lift–drag ratio, eqn. (3.26) can be differentiated with respect to α_m to give the optimum mean flow angle for maximum efficiency. Thus,

$$\frac{\partial \eta_D}{\partial \alpha_m} = \frac{4C_D \cos 2\alpha_m}{C_L \sin^2 2\alpha_m} = 0,$$

so that

$$\alpha_{m \text{ opt}} = 45 \text{ deg},$$

therefore

$$\eta_{D \text{ max}} = 1 - \frac{2C_D}{C_L}. \tag{3.27}$$

This simple analysis suggests that maximum efficiency of a compressor cascade is obtained when the mean flow angle is 45 deg, but ignores changes in the ratio C_D/C_L with varying α_m. Howell[3] has calculated the effect of having a specified variation of C_D/C_L upon cascade efficiency, comparing it with the case when C_D/C_L is constant. Figure 3.6 shows the results of this calculation as well as the variation of C_D/C_L with α_m. The graph shows that $\eta_{D \text{max}}$ is at an optimum angle only a little less than 45 deg but that the curve is rather flat for a rather wide change in α_m. Howell suggested that a value of α_m *rather less* than the optimum could well be chosen with little sacrifice in efficiency, and with some benefit with regard to power–weight ratio of compressors. In Howell's calculations, the drag is an estimate based on cascade

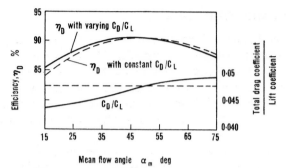

FIG. 3.6. Efficiency variation with average flow angle (adapted from Howell[3]).

experimental data together with an allowance for wall boundary-layer losses and "secondary-flow" losses.

PERFORMANCE OF TWO-DIMENSIONAL CASCADES

From the relationships developed earlier in this chapter it is apparent that the effects of a cascade may be completely deduced if the flow angles at inlet and outlet together with the pressure loss coefficient are known. However, for a given cascade only one of these quantities may be arbitrarily specified, the other two being fixed by the cascade geometry and, to a lesser extent, by the Mach number and Reynolds number of the flow. For a given family of geometrically similar cascades the performance may be expressed functionally as,

$$\zeta, a_2 = (a_1, M_1, Re), \tag{3.28}$$

where ζ is the pressure loss coefficient, eqn. (3.7), M_1 is the inlet Mach number $= c_1/(\gamma R T_1)^{\frac{1}{2}}$, Re is the inlet Reynolds number $= \rho_1 c_1 l/\mu$ based on blade chord length.

Despite numerous attempts it has not been found possible to determine, accurately, cascade performance characteristics by theoretical means alone and the experimental method still remains the most reliable technique. An account of the theoretical approach to the problem lies outside the scope of this book, however, a useful summary of the subject is given by Horlock.[4]

THE CASCADE WIND TUNNEL

The basis of much turbomachinery research and development derives from the cascade wind tunnel, e.g. Fig. 3.1 (or one of its numerous variants), and a brief description of the basic aerodynamic design is given below. A more complete description of the cascade tunnel is given by Carter *et al.*[5] including many of the research techniques developed or adopted by the National Gas Turbine Establishment at Farnborough, England.

In a well-designed cascade tunnel it is most important that the flow near the central region of the cascade blades (where the flow measurements are made) is approximately two-dimensional. This effect could be achieved by employing a large number of long blades, but an excessive amount of power would be required to operate the tunnel. With a tunnel of more reasonable size, aerodynamic difficulties become apparent and arise from the tunnel wall boundary layers interacting with the blades. In particular, and as illustrated in Fig. 3.7a, the tunnel

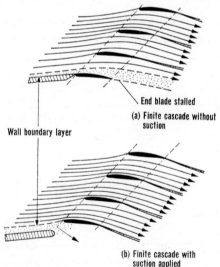

End blade stalled

(a) Finite cascade without suction

Wall boundary layer

(b) Finite cascade with suction applied

Fig. 3.7. Streamline flow through cascades (adapted from Carter[5]).

wall boundary layer mingles with the end blade boundary layer and, as a consequence, this blade stalls resulting in a non-uniform flow field.

Stalling of the end blade may be delayed by applying a controlled amount of suction to a slit just upstream of the blade, and sufficient to remove the tunnel wall boundary layer (Fig. 3.7b). Without such boundary-layer removal the effects of flow interference can be quite pronounced. They become most pronounced near the cascade "stalling point" (defined later) when any small disturbance of the upstream flow field precipitates stall on blades adjacent to the end blade. Instability of this type has been observed in compressor cascades and can affect every blade of the cascade. It is usually characterised by regular, periodic "cells" of stall crossing rapidly from blade to blade; the term *propagating stall* is often applied to the phenomenon. Some discussion of the mechanism of propagating stall is given in Chapter 6.

The boundary layers on the walls to which the blade roots are attached, generate *secondary vorticity* in passing through the blades which may produce substantial *secondary flows*. The mechanism of this phenomenon has been discussed at some length by Carter,[6] Horlock[4] and many others and a brief explanation is included in Chapter 6.

In a compressor cascade the rapid increase in pressure across the blades causes a marked thickening of the wall boundary layers and produces an effective contraction of the flow, as depicted in Fig. 3.8. A

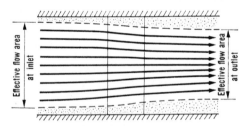

FIG. 3.8. Contraction of streamlines due to boundary layer thickening (adapted from Carter[5]).

contraction coefficient, used as a measure of the boundary-layer growth through the cascade, is defined by $\rho_1 c_1 \cos \alpha_1 / (\rho_2 c_2 \cos \alpha_2)$. Carter[5] quotes values of 0·9 for a good tunnel dropping to 0·8 in normal high-speed tunnels and even less in bad cases. These are values for compressor cascades; with turbine cascades slightly higher values can be expected.

Because of the contraction of the main through flow, the theoretical pressure rise across a compressor cascade, even allowing for losses, is never achieved. This will be evident since a contraction (in a subsonic flow) accelerates the fluid, which is in conflict with the diffuser action of the cascade.

To counteract these effects it is customary (in Great Britain) to use *at least* seven blades in a compressor cascade, each blade having a minimum aspect ratio (blade span–chord length) of 3. With seven blades, suction is desirable in a compressor cascade but it is not usual in a turbine cascade. In the United States much lower aspect ratios are commonly employed in compressor cascade testing, the technique being the almost complete removal of tunnel wall boundary layers from all four walls using a combination of suction slots and perforated end walls to which suction is applied.

CASCADE TEST RESULTS

The basic cascade performance data are obtained from measurements of pressure, flow angle and velocity taken across one or more pitches at entry to and at exit from the cascade. Some details of the instrumentation used for these test measurements are given in refs. 4, 5 and 7. One of the most useful works published on methods for deriving wind speed and flow direction from measurements of pressure in the flow is the National Physical Laboratory publication, ref. 7. This contains details of the design and construction of various instruments as well as their general principles and practical performance. An extensive bibliography on all types of measurement in fluid flow is given by Dowden.[18] Figure 3.9 shows a typical cascade test result from a traverse across 2 blade pitches taken by Todd,[8] at an inlet Mach number of 0·6. It is observed that a total pressure deficit occurs across the blade row arising from the fluid friction on the blades. The fluid deflection is not uniform and is a maximum at each blade trailing edge on the pressure side of the blades. From such test results, average values of total pressure loss and fluid outlet angle are found (usually on a mass flow basis). The use of terms like "total pressure loss" and "fluid outlet angle" in the subsequent discussion will signify these *average* values.

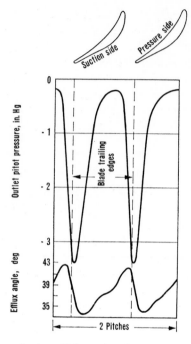

FIG. 3.9. A sample plot of inlet and outlet stagnation pressures and
fluid outlet angle (adapted from Todd[8]).

Similar tests performed for a range of fluid inlet angles, at the same
inlet Mach number M_1 and Reynolds number Re, enables the complete
performance of the cascade to be determined (at that M_1 and Re). So
as to minimise the amount of testing required, much cascade work is
performed at low inlet velocities, but at a Reynolds number greater
than the "critical" value. This critical Reynolds number Re_c is approximately 2×10^5 based on inlet velocity and blade chord. With $Re > Re_c$,
total pressure losses and fluid deflections are only slightly dependent on
changes in Re. Mach number effects are negligible when $M_1 < 0.3$.
Thus, the performance laws, eqn. (3.28), for this flow simplify to,

$$\zeta, a_2 = f(a_1). \tag{3.28a}$$

There is a fundamental difference between the flows in turbine cascades and those in compressor cascades which needs emphasising. A fluid flowing through a channel in which the mean pressure is falling (mean flow is accelerating) experiences a relatively small total pressure loss in contrast with the mean flow through a channel in which the pressure is rising (diffusing flow) when losses may be high. This characteristic difference in flow is reflected in turbine cascades by a wide range of low loss performance and in compressor cascades by a rather narrow range.

COMPRESSOR CASCADE PERFORMANCE

A typical set of low-speed compressor cascade results,[9] for a blade cascade of specified geometry, is shown in Fig. 3.10. These results are presented in the form of a pressure loss coefficient $\Delta p_0/(\frac{1}{2}\rho c_1^2)$ and fluid deflection $\epsilon = a_1 - a_2$ against incidence $i = a_1 - a_1'$ (refer to Fig. 3.2 for nomenclature). Note that from eqn. (3.7), $\Delta p_0/(\frac{1}{2}\rho c_1^2) = \zeta \cos^2 a_1$. There is a pronounced increase in total pressure loss as the incidence rises beyond a certain value and the cascade is *stalled* in this region. The precise incidence at which stalling occurs is difficult to define and a

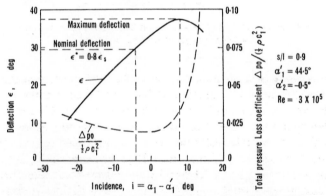

FIG. 3.10. Compressor cascade characteristics (Howell[9]). (By courtesy of the Controller of H.M.S.O., Crown copyright reserved.)

stall point is arbitrarily specified as the incidence at which the total pressure loss is *twice* the minimum loss in total pressure. Physically,

stall is characterised (at positive incidence) by the flow separating from
the suction side of the blade surfaces. With decreasing incidence, total
pressure losses again rise and a "negative incidence" stall point can also
be defined as above. The *working range* is conventionally defined as the
incidence range between these two limits at which the losses are twice
the minimum loss. Accurate knowledge of the extent of the working
range, obtained from two-dimensional cascade tests, is of great importance when attempting to assess the suitability of blading for changing
conditions of operation. A *reference incidence* angle can be most
conveniently defined either at the mid-point of the working range or,
less precisely, at the minimum loss condition. These two conditions do
not necessarily give the same reference incidence.

From such cascade test results the *profile losses* through compressor
blading of the same geometry may be estimated. To these losses
estimates of the annulus skin friction losses and other secondary losses
must be added, and from which the efficiency of the compressor blade
row may be determined. Howell[10] suggested that these losses could be
estimated using the following drag coefficients. For the annulus walls
loss,

$$C_{Da} = 0 \cdot 02 \, s/H \qquad (3.29a)$$

and for the so-called "secondary" loss,

$$C_{Ds} = 0 \cdot 018 \, C_L^2 \qquad (3.29b)$$

where s, H are the blade pitch and blade length respectively, and C_L the
blade lift coefficient. Calculations of this type were made by Howell
and others to estimate the efficiency of a complete compressor stage. A
worked example to illustrate the details of the method is given in
Chapter 5. Figure 3.11 shows the variation of stage efficiency with flow
coefficient and it is of particular interest to note the relative magnitude
of the profile losses in comparison with the overall losses, especially at
the design point.

Cascade performance data to be easily used, are best presented in
some condensed form. Several methods of empirically correlating low-
speed performance data have been developed in Great Britain. Howell's
correlation[9] relates the performance of a cascade to its performance at

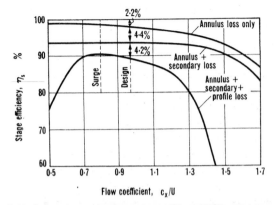

FIG. 3.11. Losses in a compressor stage (Howell[10]). (Courtesy of the Institution of Mechanical Engineers.)

a "nominal" condition defined at 80% of the stalling deflection. Carter[11] has referred performance to an optimum incidence given by the highest lift–drag ratio of the cascade. In the United States, the National Advisory Committee for Aeronautics (NACA), now called the National Aeronautics and Space Administration (NASA), systematically tested whole families of different cascade geometries, in particular, the widely used NACA 65 Series.[12] The data on the NACA 65 Series has been usefully summarised by Felix[13] where the performance of a fixed geometry cascade can be more readily found. A concise summary is also given by Horlock.[4]

TURBINE CASCADE PERFORMANCE

Figure 3.12 shows results obtained by Ainley[14] from two sets of turbine cascade blades, impulse and "reaction". The term "reaction" is used here to denote, in a qualitative sense, that the fluid accelerates through the blade row and thus experiences a *pressure drop* during its passage. There is no pressure change across an impulse blade row. The performance is expressed in the form $\lambda = \Delta p_0/(p_{02} - p_2)$ and α_2 against incidence.

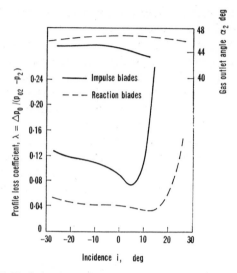

Fig. 3.12. Variation in profile loss with incidence for typical turbine
blades (adapted from Ainley[14]).

From these results it is observed that:

(a) the reaction blades have a much wider range of low loss performance than the impulse blades, a result to be expected as the blade boundary layers are subjected to a favourable pressure gradient,

(b) the fluid outlet angle α_2 remains relatively constant over the whole range of incidence in contrast with the compressor cascade results.

For turbine cascade blades, a method of correlation is given by Ainley[15] which enables the performance of a gas turbine to be predicted with an estimated tolerance of within 2% on peak efficiency. In Chapter 4 a rather different approach, using a method attributed to Soderberg, is outlined. While being possibly slightly less accurate than Ainley's correlation, Soderberg's method employs fewer parameters and is rather easier to apply.

COMPRESSOR CASCADE CORRELATIONS

Many experimental investigations have confirmed that the efficient performance of compressor cascade blades is limited by the growth and separation of the blade surface boundary layers. One of the aims of cascade research is to establish the generalised loss characteristics and stall limits of conventional blades. This task is made difficult because of the large number of factors which can influence the growth of the blade surface boundary layers, viz. surface velocity distribution, blade Reynolds number, inlet Mach number, free-stream turbulence and unsteadiness, and surface roughness. From the analysis of experimental data several correlation methods have been evolved which enable the first-order behaviour of the blade losses and limiting fluid deflection to be predicted with sufficient accuracy for engineering purposes.

LIEBLEIN. The correlation of Lieblein[17,19] is based on the experimental observation that a large amount of velocity diffusion on blade surfaces tends to produce thick boundary layers and eventual flow

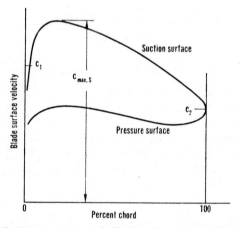

FIG. 3.13. Compressor cascade blade surface velocity distribution.

separation. Lieblein states the general hypothesis that in the region of minimum loss, the wake thickness and consequently the magnitude of the loss in total pressure, is proportional to the diffusion in velocity on the suction-surface of the blade in that region. The hypothesis is based on the consideration that the boundary layer on the suction-surface of conventional compressor blades contributes the largest share of the blade wake. Therefore, the suction-surface velocity distribution becomes the main factor in determining the total pressure loss.

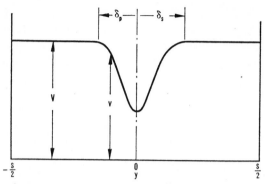

Fig. 3.14. Model variation in velocity in a plane normal to axial direction.

Figure 3.13 shows a typical velocity distribution derived from surface pressure measurements on a compressor cascade blade in the region of minimum loss. The diffusion in velocity may be expressed as the ratio of maximum suction-surface velocity to outlet velocity, $c_{max,s}/c_2$. Lieblein found a correlation between the diffusion ratio $c_{max,s}/c_2$ and the wake momentum-thickness to chord ratio, θ_2/l at the reference incidence (mid-point of working range) for American NACA 65-(A_{10}) and British C.4 circular-arc blades. The wake momentum-thickness, with the parameters of the flow model in Fig. 3.14 is defined as

$$\theta_2 = \int_{\delta_p}^{\delta_s} \frac{v}{V}\left(1 - \frac{v}{V}\right) dy. \tag{3.30}$$

The Lieblein correlation, with his data points removed for clarity, is closely fitted by the mean curve in Fig. 3.15. This curve represents the equation

$$\frac{\theta_2}{l} = 0 \cdot 004 \Big/ \left\{1 - 1 \cdot 17 \ln\left(\frac{c_{max,s}}{c_2}\right)\right\} \tag{3.31}$$

which may be more convenient to use in calculating results. It will be noticed that for the limiting case when $(\theta_2/l) \to \infty$, the corresponding *upper* limit for the diffusion ratio $c_{max,s}/c_2$ is 2·35. The *practical* limit of

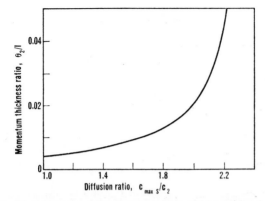

FIG. 3.15. Mean variation of wake momentum-thickness/chord ratio with suction-surface diffusion ratio at reference incidence condition for NACA 65-($C_{10}A_{10}$)10 blades and British C.4 circular-arc blades (adapted from Lieblein[19]).

efficient operation would correspond to a diffusion ratio of between 1·9 and 2·0.

Losses are usually expressed in terms of the stagnation pressure loss coefficient $\bar{\omega} = \Delta p_0/(\tfrac{1}{2}\rho c_1^2)$ or $\zeta = \Delta p_0/(\tfrac{1}{2}\rho c_x^2)$ as well as the drag coefficient C_D. Lieblein and Roudebush[20] have demonstrated the simplified relationship between momentum-thickness ratio and total pressure loss coefficient, valid for unstalled blades,

$$\bar{\omega} = 2\left(\frac{\theta_2}{l}\right)\left(\frac{l}{s}\right)\frac{\cos^2 \alpha_1}{\cos^3 \alpha_2}. \qquad (3.32)$$

Combining this relation with eqns. (3.7) and (3.17) the following useful results can be obtained:

$$C_D = \bar{\omega}\left(\frac{s}{l}\right)\frac{\cos^3 \alpha_m}{\cos^2 \alpha_1} = 2\left(\frac{\theta_2}{l}\right)\left(\frac{\cos \alpha_m}{\cos \alpha_2}\right)^3 = \zeta\left(\frac{s}{l}\right)\cos^3 \alpha_m. \qquad (3.33)$$

The correlation given above *assumes* a knowledge of suction-surface velocities in order that total pressure loss and stall limits can be estimated. As this data may be unavailable it is necessary to establish an *equivalent diffusion ratio*, approximately equal to $c_{\max,s}/c_2$, that can be easily calculated from the inlet and outlet conditions of the cascade. An empirical correlation was established by Lieblein[19] between a circulation parameter defined by $f(\Gamma) = \Gamma \cos \alpha_1/(lc_1)$ and $c_{max,s}/c_1$ at the reference incidence, where the ideal circulation $\Gamma = s(c_{y1} - c_{y2})$, using eqn. (3.23). The correlation obtained is the simple *linear* relation.

$$c_{max,s}/c_1 = 1{\cdot}12 + 0{\cdot}61f(\Gamma) \tag{3.34}$$

which applies to both NACA 65-(A_{10}) and C.4 circular arc blades. Hence, the equivalent diffusion ratio, after substituting for Γ and simplifying, is

$$D_{eq} = \frac{c_{max,s}}{c_2} = \frac{\cos \alpha_2}{\cos \alpha_1} \left\{ 1{\cdot}12 + 0{\cdot}61 \left(\frac{s}{l} \right) \cos^2 \alpha_1 (\tan \alpha_1 - \tan \alpha_2) \right\}.$$

$$\tag{3.35}$$

At incidence angles greater than reference incidence, Lieblein found that the following correlation was adequate:

$$D_{eq} = \frac{\cos \alpha_2}{\cos \alpha_1} \left\{ 1{\cdot}12 + k(i - i_{ref})^{1{\cdot}43} + 0{\cdot}61 \left(\frac{s}{l} \right) \cos^2 \alpha_1 (\tan \alpha_1 - \tan \alpha_2) \right\}$$

$$\tag{3.36}$$

where $k = 0{\cdot}0117$ for the NACA 65-(A_{10}) blades and $k = 0{\cdot}007$ for the C.4 circular arc blades.

The expressions given above are still very widely used as a means of estimating total pressure loss and the unstalled range of operation of blades commonly employed in subsonic axial compressors. The method has been modified and extended by Swann[21] to include the additional losses caused by shock waves in transonic compressors. The discussion of transonic compressors is outside the scope of this text and is not included.

HOWELL. The low-speed correlation of Howell[9] has been widely used by designers of axial compressors and is based on a nominal condition such that the deflection ε^* is 80% of the stalling deflection, ε_s (Fig. 3.10). Choosing $\varepsilon^* = 0{\cdot}8\varepsilon_s$ as the *design condition* represents a compromise between the ultraconservative and the overoptimistic! Howell found that the nominal deflections of various compressor

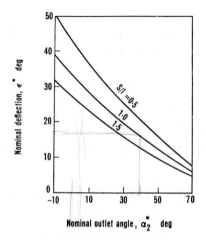

FIG. 3.16. Variation of nominal deflection with nominal outlet angle for several space/chord ratios (adapted from Howell[10]).

cascades are, primarily, a function of the space-chord ratio s/l, the nominal fluid outlet angle a_2^* and the Reynolds number Re

$$\varepsilon^* = f(s/l, a_2^*, Re). \tag{3.37}$$

It is important to note that the correlation (which is really a correlation of stalling deflection, $\varepsilon_s = 1{\cdot}25\varepsilon^*$) is virtually independent of blade camber θ in the normal range of choice of this parameter ($20° < \theta < 40°$). Figure 3.16 shows the variation of ε^* found by Howell,[10] against a_2^* for several space–chord ratios. The dependence on Reynolds number is small for $Re > 3 \times 10^5$, based on blade chord.

An approximating formula to the data given in Fig. 3.16, which was quoted by Howell and frequently found to be useful in preliminary performance estimation, is the tangent-difference rule:

$$\tan a_1^* - \tan a_2^* = \frac{1{\cdot}55}{1 + 1{\cdot}5\, s/l} \tag{3.38}$$

which is applicable in the range $0 \leqq a_2^* \leqq 40°$.

FLUID DEVIATION

The *difference* between the fluid and blade inlet angles at cascade inlet is under the arbitrary control of the designer. At cascade outlet however, the difference between the fluid and blade angles, called the *deviation* δ, is a function of blade camber, blade shape, space–chord ratio and stagger angle. Referring to Fig. 3.2, the deviation $\delta = a_2 - a_2'$ is drawn as positive; almost without exception it is in such a direction that the deflection of the fluid is reduced. The deviation may be of considerable magnitude and it is important that an accurate estimate is made of it. Re-examining Fig. 3.9 again, it will be observed that the fluid receives its maximum guidance on the pressure side of the cascade channel and that this diminishes almost linearly towards the suction side of the channel.

Howell used an empirical rule to relate nominal deviation δ^* to the camber and space–chord ratio,

$$\delta^* = m\theta(s/l)^n, \tag{3.39}$$

where $n \doteq \tfrac{1}{2}$ for compressor cascades and $n \doteq 1$ for compressor *inlet guide vanes*. The value of m depends upon the shape of the camber line and the blade setting. For a compressor cascade (i.e. diffusing flow),

$$m = 0.23(2a/l)^2 + a_2^*/500, \tag{3.40a}$$

where a is the distance of maximum camber from the leading edge. For the inlet guide vanes, which are essentially *turbine* nozzles (i.e. accelerating flow),

$$m = \text{constant} = 0.19 \tag{3.40b}$$

ILLUSTRATIVE EXAMPLE: A compressor cascade has a space–chord ratio of unity and blade inlet and outlet angles of 50 deg and 20 deg respectively. If the blade camber line is a circular arc (i.e. $a/l = 50\%$) and the cascade is designed to operate at Howell's nominal condition, determine the fluid deflection, incidence and ideal lift coefficient at the design point.

The camber, $\theta = a_1' - a_2' = 30$ deg. As a first approximation put $a_2^* = 20$ deg in eqn. (3.40) to give $m = 0.27$ and, using eqn. (3.39), $\delta^* = 0.27 \times 30 = 8.1$ deg. As a better approximation put $a_2^* = 28.1$ deg in eqn. (3.40) giving $m = 0.2862$ and $\delta^* = 8.6$ deg. Thus, $a_2^* = 28.6$ deg is sufficiently accurate.

From Fig. 3.16, with $s/l = 1.0$ and $a_2^* = 28.6$ deg obtain $\varepsilon^* = a_1^* - a_2^* = 21$ deg. Hence $a_1^* = 49.6$ deg and the nominal incidence $i^* = a_1^* - a_1'$

i ≑ 0·4
(like negative angle of attack)

$= -0.4$ deg.

The *ideal* lift coefficient is found by setting $C_D = 0$ in eqn. (3.18),

$$C_L = 2(s/l) \cos a_m(\tan a_1 - \tan a_2).$$

Putting $a_1 = a_1^*$, $a_2 = a_2^*$ and noting $\tan a_m^* = \frac{1}{2}(\tan a_1^* + \tan a_2^*)$ obtain $a_m^* = 40.75$ deg and $C_L^* = 2(1.172 - 0.545)0.758 \doteqdot 0.95.$

In conclusion it will be noted that the estimated deviation is one of the most important quantities for design purposes, as small errors in it are reflected in large changes in deflection and thus, in predicted performance.

OFF-DESIGN PERFORMANCE

To obtain the performance of a given cascade at conditions removed from the design point, generalised performance curves of Howell[9] shown in Fig. 3.17 may be used. If the nominal deflection ε^* and nominal incidence i^* are known the off-design performance (deflection, total pressure loss coefficient) of the cascade at any other incidence is readily calculated.

EXAMPLE: In the previous exercise, with a cascade of $s/l = 1.0$, $a_1' = 50$ deg and $a'_2 = 20$ deg the nominal conditions were $\varepsilon^* = 21$ deg and $i^* = -0.4$ deg. Suppose the off-design performance of this cascade is required at an incidence $i = 3.8$ deg. Referring to Fig. 3.17 and with $(i-i^*)/\varepsilon^* = 0.2$ obtain $C_D \doteqdot 0.017$, $\varepsilon/\varepsilon^* = 1.15$. Thus, the off-design deflection, $\varepsilon = 24.1$ deg.

From eqn. (3.17), the total pressure loss coefficient is,

$$\zeta = \Delta p_0/(\tfrac{1}{2}\rho c_x^2) = C_D/[(s/l)\cos^3 a_m].$$

Now $a_1 = a_1' + i = 53.8$ deg, also $a_2 = a_1 - \varepsilon = 29.7$ deg, therefore,

$$a_m = \tan^{-1}\{\tfrac{1}{2}(\tan a_1 + \tan a_2)\} = \tan^{-1}\{0.969\} = 44.1 \text{ deg,}$$

hence

$$\zeta = 0.017/0.719^3 = 0.0458.$$

The tangential lift force coefficient, eqn. (3.9), is

$$C_f = (p_2 - p_1)/(\tfrac{1}{2}\rho c_x^2) = 2(\tan a_1 - \tan a_2) = 1.596.$$

The diffuser efficiency, eqn. (3.25), is

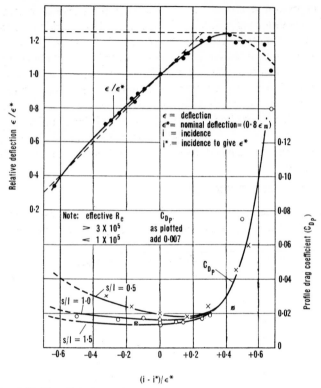

FIG. 3.17. The off-design performance of a compressor cascade (Howell[9]). (By courtesy of the Controller of H.M.S.O., Crown copyright reserved.)

$$\eta_D = 1 - \zeta/(C_f \tan \alpha_m) = 1 - 0.0458/(1.596 \times 0.969) = 97\%.$$

It is worth noting, from the representative data contained in the above exercise, that the validity of the approximation in eqn. (3.20) is amply justified.

Howell's correlation, clearly, is a simple and fairly direct method of assessing the performance of a given cascade for a range of inlet flow angles. The data can also be used for solving the more complex *inverse problem*, namely, the selection of a suitable cascade geometry when the fluid deflection is given. For this case, if the previous method of a nominal design condition is used, mechanically unsuitable space–chord ratios are a possibility. The space–chord ratio may, however, be determined to some extent by the mechanical layout of the compressor, the design incidence then only fortuitously coinciding with the nominal incidence. The design incidence is therefore somewhat arbitrary and some designers, ignoring nominal design conditions, may select an incidence best suited to the operating conditions under which the compressor will run. For instance, a negative design incidence may be chosen so that at reduced flow rates a *positive* incidence condition is approached.

MACH NUMBER EFFECTS

High-speed cascade characteristics are similar to those at low speed until the *critical* Mach number M_c is reached, after which the performance declines. Figure 3.18, taken from Howell,[9] illustrates for a particular cascade tested at varying Mach number and fixed incidence, the drastic decline in pressure rise coefficient up to the *maximum* Mach

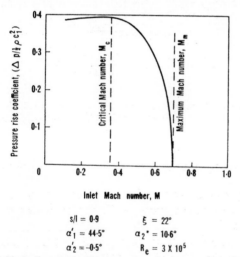

$$s/l = 0.9 \qquad \xi = 22°$$
$$\alpha_1' = 44.5° \qquad \alpha_2^* = 10.6°$$
$$\alpha_2' = -0.5° \qquad R_e = 3 \times 10^5$$

FIG. 3.18. Variation of cascade pressure rise coefficient with inlet Mach number (Howell[9]). (By courtesy of the Controller of H.M.S.O., Crown copyright reserved.)

number at entry M_m, when the cascade is fully choked. When the cascade is choked, no further increase in mass flow through the cascade is possible. The definition of inlet critical Mach number is less precise, but one fairly satisfactory definition[4] in use is that the maximum *local* Mach number in the cascade has reached unity.

Howell attempted to correlate the decrease in both efficiency and deflection in the range of inlet Mach numbers, $M_c \leq M \leq M_m$ and these are shown in Fig. 3.19. By employing this correlation, curves similar to that in Fig. 3.18 may be found for each incidence.

One of the principal aims of high-speed cascade testing is to obtain data for determining the values of M_c and M_m. Howell[10] indicates how, for a typical cascade, M_c and M_m vary with incidence (Fig. 3.20).

TURBINE CASCADE CORRELATION (Ainley)

In 1951, Ainley and Mathieson[15] published a method of estimating the performance of an axial flow turbine and the method has been widely used ever since. In essence the method determines the total pressure loss and gas efflux angle for each row of a turbine stage at a

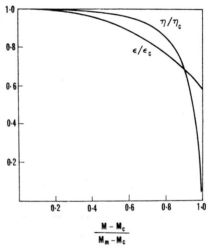

FIG. 3.19. Variation of efficiency and deflection with Mach number (adapted from Howell[9]).

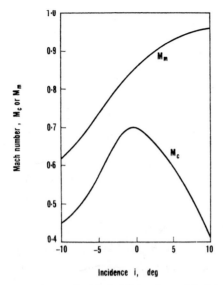

FIG. 3.20. Dependence of critical and maximum Mach numbers upon incidence (Howell[10]). (By courtesy of the Institution of Mechanical Engineers.)

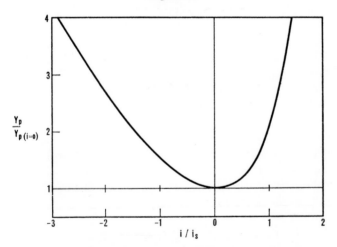

FIG. 3.21. Variation of profile loss with incidence for typical turbine blading (adapted from Ainley and Mathieson[15]).

single *reference* diameter and under a wide range of inlet conditions. This reference diameter was taken as the arithmetic mean of the rotor and stator rows inner and outer diameters. More recently Dunham and Came[22] have gathered together details of several improvements to the method of Ainley and Mathieson which give better performance prediction for *small* turbines than did the original method. When the blading is *competently* designed the revised method appears to give reliable predictions of efficiency to within 2% over a wide range of designs, sizes and operating conditions.

Total pressure loss correlations

The overall total pressure loss is composed of three parts, viz. (i) profile loss, (ii) secondary loss, and (iii) tip clearance loss.

(i) A profile loss coefficient is defined as the loss in stagnation pressure across the blade row or cascade, divided by the difference between stagnation and static pressures at blade outlet; i.e.

$$Y_p = \frac{p_{01} - p_{02}}{p_{02} - p_2} . \qquad (3.41)$$

In the Ainley method, profile loss is determined initially at zero incidence ($i = 0$). At any other incidence the profile loss ratio $Y_p/Y_{p(i=0)}$ is assumed to be defined by a unique function of the incidence ratio i/i_s (Fig. 3.21), where i_s is the stalling incidence. This is defined as the incidence at which $Y_p/Y_{p(i=0)} = 2 \cdot 0$.

Ainley and Mathieson correlated the profile losses of turbine blade rows against space/chord ratio s/l, fluid outlet angle α_2, blade maximum thickness/chord ratio t/l and blade inlet angle. The variation of $Y_{p(l=0)}$ against s/l is shown in Fig. 3.22 for nozzles and impulse blading at various flow outlet angles. The sign convention used for flow angles in a turbine cascade is indicated in Fig. 3.24. For other types of blading intermediate between nozzle blades and impulse blades the following expression is employed:

$$Y_{p(i=0)} = \left\{ Y_{p(\alpha 1 = 0)} + \left(\frac{\alpha_1}{\alpha_2}\right)^2 \left[Y_{p(\alpha 1 = \alpha 2)} - Y_{p(\alpha 1 = 0)} \right] \right\} \left(\frac{t/l}{0 \cdot 2}\right)^{\alpha_1/\alpha_2}$$

$$(3.42)$$

where all the Y_p's are taken at the same space/chord ratio and flow outlet angle. If rotor blades are being considered, put β_2 for α_1 and β_3 for α_2. Equation (3.42) includes a correction for the effect of thickness–chord ratio and is valid in the range $0.15 \leq t/l \leq 0.25$. If the actual blade has a t/l greater or less than the limits quoted, Ainley recommends that the loss should be taken as equal to a blade having t/l either 0.25 or 0.15. By substituting $\alpha_1 = \alpha_2$ and $t/l = 0.2$ in eqn. (3.42), the zero incidence loss coefficient for the impulse blades $Y_{p(\alpha1=\alpha2)}$ given in Fig. 3.22 is recovered. Similarly, with $\alpha_1 = 0$ at $t/l = 0.2$ in eqn. (3.42) gives $Y_{p(\alpha1=0)}$ of Fig. 3.22.

A feature of the losses given in Fig. 3.22 is that, compared with the impulse blades, the nozzle blades have a much lower loss coefficient. This trend confirms the results shown in Fig. 3.12, that flow in which the mean pressure is falling has a lower loss coefficient than a flow in which the mean pressure is constant or increasing.

(ii) The secondary losses arise from complex three-dimensional flows set up as a result of the end wall boundary layers passing through the cascade. There is substantial evidence that the end wall boundary layers are convected inwards along the suction-surface of the blades as the main flow passes through the blade row, resulting in a serious maldistribution of the flow, with losses in stagnation pressure often a significant fraction of the total loss. Ainley found that secondary losses could be represented by

$$C_{Ds} = \lambda C_L^2/(s/l) \qquad (3.43)$$

where λ is a parameter which is a function of the flow acceleration through the blade row. From eqn. (3.17), together with the definition of Y, eqn. (3.41) for incompressible flow, $C_D = Y(s/l) \cos^3 \alpha_m/\cos^2 \alpha_2$, hence

$$Y_s = \frac{C_{Ds} \cos^2 \alpha_2}{(s/l) \cos^3 \alpha_m} = \lambda \left(\frac{C_L}{s/l}\right)^2 \frac{\cos^2 \alpha_2}{\cos^3 \alpha_m} = \lambda Z \qquad (3.44)$$

where Z is the blade aerodynamic loading coefficient. Dunham[23] subsequently found that this equation was not correct for blades of low aspect ratio, as in small turbines. He modified Ainley's result to include a better correlation with aspect ratio and at the same time simplified the

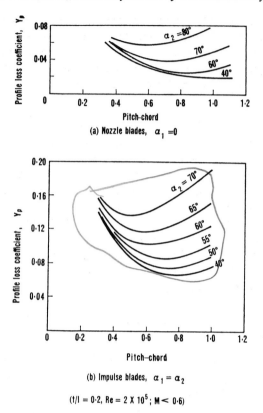

FIG. 3.22. Profile loss coefficients of turbine nozzle and impulse blades at zero incidence ($t/l = 20\%$; $Re = 2 \times 10^5$; $M < 0.6$) (adapted from Ainley and Mathieson[15]).

flow acceleration parameter. The correlation, given by Dunham and Came,[22] is

$$Y_s = 0.0334 \left(\frac{l}{H}\right) \left(\frac{\cos \alpha_2}{\cos \alpha_1}\right) Z \qquad (3.45)$$

and this represents a significant improvement in the prediction of secondary losses using Ainley's method.

Recently, more advanced methods of predicting losses in turbine blade rows have been suggested which take into account the *thickness* of the entering boundary layers on the annulus walls. Came[24] has measured the secondary flow losses on *one* end wall of several turbine cascades for various thicknesses of inlet boundary layer. He correlated his own results, and those of several other investigators, and obtained a modified form of Dunham's earlier result, viz.,

$$Y_s = \left(0.25\, Y_1 \frac{\cos^2 \alpha_1}{\cos^2 \alpha_2} + 0.009 \frac{l}{H}\right)\left(\frac{\cos \alpha_2}{\cos \alpha_1}\right) Z - Y_1 \qquad (3.46)$$

which is the net secondary loss coefficient for *one* end wall only and where Y_1 is a mass-averaged inlet boundary layer total pressure loss coefficient. It is evident that the increased accuracy obtained by use of eqn. (3.46) requires the additional effort of calculating the wall boundary layer development. In *initial* calculations of performance it is probably sufficient to use the earlier result of Dunham and Came, eqn. (3.45), to achieve a reasonably accurate result.

(iii) The tip clearance loss coefficient Y_k depends upon the blade loading Z and the size and nature of the clearance gap k. Dunham and Came presented an amended version of Ainley's original result for Y_k:

$$Y_k = B\left(\frac{l}{H}\right)\left(\frac{k}{l}\right)^{0.78} Z \qquad (3.47)$$

where $B = 0.5$ for a plain tip clearance, 0.25 for shrouded tips.

Reynolds number correction

Ainley and Mathieson[15] obtained their data for a mean Reynolds number of 2×10^5 based on the mean chord and exit flow conditions from the turbine state. They recommended for lower Reynolds numbers, down to 5×10^4, that a correction be made to stage efficiency according to the rough rule:

$$(1 - \eta_{tt}) \propto Re^{-1/5}.$$

Dunham and Came[22] gave an optional correction which is applied directly to the sum of the profile and secondary loss coefficients for a blade row using the Reynolds number *appropriate* to that row. The rule is:

$$Y_p + Y_s \propto Re^{-1/5}.$$

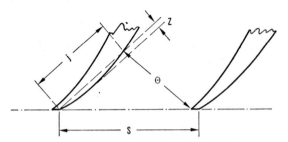

FIG. 3.23. Details near turbine cascade exit showing "throat" and suction-surface curvature parameters.

Flow outlet angle from a turbine cascade

It was pointed out by Ainley[14] that the method of defining deviation angle as adopted in several well-known compressor cascade correlations had proved to be impracticable for turbine blade cascade. In order to predict fluid outlet angle a_2, steam turbine designers had made much use of the simple empirical rule that

$$a_2 = \cos^{-1} \Theta/s \qquad (3.48\text{a})$$

where Θ is the opening at the throat, depicted in Fig. 3.23, and s is the pitch. This widely used rule gives a very good approximation to measured pitchwise averaged flow angles when the outlet Mach number is at or close to unity. However, at low Mach numbers substantial variations have been found between the rule and observed flow angles. Ainley and Mathieson[15] recommended that for low outlet Mach numbers $0 < M_2 \leqq 0\cdot5$, the following rule be used:

$$a_2 = f(\cos^{-1}\Theta/s) + 4s/e \text{ (deg)} \qquad (3.48\text{b})$$

where $f(\cos^{-1}\Theta/s) = -11\cdot15 + 1\cdot154 \cos^{-1}\Theta/s$ and $e = j^2/(8z)$ is the mean radius of curvature of the blade suction surface between the throat and the trailing edge. At a gas outlet Mach number of unity Ainley and Mathieson assumed, for a turbine blade row, that

$$a_2 = \cos^{-1} A_t/A_{n2} \qquad (3.48\text{c})$$

where A_t is the passage throat area and A_{n2} is the annulus area in the reference plane downstream of the blades. If the annulus walls at the ends of the cascade are not flared then eqn. (3.48c) is the same as eqn. (3.48a). Between $M_2 = 0.5$ and $M_2 = 1.0$ a linear variation of α_2 can be reasonably assumed in the absence of any other data.

OPTIMUM SPACE–CHORD RATIO OF TURBINE BLADES (Zweifel)

It is worth pondering a little upon the effect of the space–chord ratio in turbine blade rows as this is a factor strongly affecting efficiency. Now if the spacing between blades is made small, the fluid then tends to receive the maximum amount of guidance from the blades, but the friction losses will be very large. On the other hand, with the same blades spaced well apart, friction losses are small but, because of poor fluid guidance, the losses resulting from flow separation are high. These considerations led Zweifel[16] to formulate his criterion for the optimum space–chord ratio of blading having large deflection angles. Essentially, *Zweifel's criterion* is simply that the ratio (ψ_T) of the actual to an "ideal" tangential blade loading, has a certain constant value for minimum losses. The tangential blade loads are obtained from the real and ideal pressure distributions on both blade surfaces, as described below.

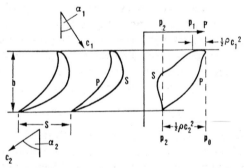

FIG. 3.24. Pressure distribution around a turbine cascade blade (after Zweifel[16]).

Figure 3.24 indicates a typical pressure distribution around one blade in a turbine cascade, curves P and S corresponding to the pressure (or concave) side and suction (convex) side respectively. The pressures are projected parallel to the cascade front so that the area enclosed between the curves S and P represents the *actual tangential blade load per unit span,*

$$Y = \rho s c_x (c_{y2} + c_{y1}), \tag{3.49}$$

cf. eqn. (3.3) for a compressor cascade.

It is instructive to examine the pressures along the blade surfaces. Assuming incompressible flow the static inlet pressure is $p_1 = p_0 - \frac{1}{2}\rho c_1^2$; if losses are also ignored the outlet static pressure $p_2 = p_0 - \frac{1}{2}\rho c_2^2$. The pressure on the P side remains high at first (p_0 being the maximum, attained only at the stagnation point), then falls sharply to p_2. On the S side there is a rapid decrease in static pressure from the leading edge, but it may even rise again towards the trailing edge. The closer the blade spacing s the smaller the load Y becomes (eqn. (3.49)). Conversely, wide spacing implies an increased load with pressure rising on the P side and falling on the S side. Now, whereas the static pressure can never rise above p_0 on the P surface, very low pressures are possible, at least in theory on the S surface. However, the pressure rise towards the trailing edge is limited in practice if flow separation is to be avoided, which implies that the load carried by the blade is restricted.

To give some idea of blade load capacity, the real pressure distribution is compared with an ideal pressure distribution giving a maximum load Y_{id} without risk of fluid separation on the S surface. Upon reflection, one sees that these conditions for the ideal load are fulfilled by p_0 acting over the *whole P* surface and p_2 acting over the *whole S* surface. With this ideal pressure distribution (which cannot, of course, be realised), the tangential load per unit span is,

$$Y_{id} = \frac{1}{2}\rho c_2^2 b \tag{3.50}$$

and, therefore,

$$\psi_T = Y/Y_{id} = 2(s/b)\cos^2 \alpha_2 (\tan \alpha_1 + \tan \alpha_2) \tag{3.51}$$

after combining eqns. (3.49) and (3.50) together with angles defined by the geometry of Fig. 3.24.

Zweifel found from a number of experiments on turbine cascades that for minimum losses the value of ψ_T was approximately 0·8. Thus, for specified inlet and outlet angles the optimum space–chord ratio can be estimated. However, according to Horlock,[25] Zweifel's criterion predicts optimum space–chord ratio for the data of Ainley only for outlet angles of 60 to 70 deg. At other outlet angles it does not give an accurate estimate of optimum space–chord ratio.

REFERENCES

1. CSANADY, G. T., *Theory of Turbomachines*. McGraw-Hill, New York (1964).
2. GLAUERT, H., *Aerofoil and Airscrew Theory*. 2nd ed. Cambridge University Press (1959).
3. HOWELL, A. R., Design of axial compressors. *Proc. Instn. Mech. Engrs. London* **153** (1945).
4. HORLOCK, J. H., *Axial Flow Compressors*. Butterworth, London (1958).
5. CARTER, A. D. S., ANDREWS, S. J. and SHAW, H., Some fluid dynamic research techniques. *Proc. Instn. Mech. Engrs. London*, **163** (1950).
6. CARTER, A. D. S., Three-dimensional flow theories for axial compressors and turbines. *Proc. Instn. Mech. Engrs. London*, **159** (1948).
7. BRYER, D. W. and PANKHURST, R. C., *Pressure-probe Methods for Determining Wind Speed and Flow Direction*. National Physical Laboratory, H.M.S.O. (1971).
8. TODD, K. W., Practical aspects of cascade wind tunnel research. *Proc. Instn. Mech. Engrs. London*, **157** (1947).
9. HOWELL, A. R., The present basis of axial flow compressor design: Part I—Cascade theory and performance. *A.R.C.R. and M.* 2095 (1942).
10. HOWELL, A. R., Fluid dynamics of axial compressors. *Proc. Instn. Mech. Engrs. London*, **153** (1945).
11. CARTER, A. D. S., Low-speed performance of related aerofoils in cascade. *A.R.C. C.P.* No. 29 (1950).
12. HERRIG, L. J., EMERY, J. C. and ERWIN, J. R., Systematic two-dimensional cascade tests of NACA 65-Series compressor blades at low speeds. *NACA T.N.* 3916 (1957).
13. FELIX, A. R., Summary of 65-Series compressor blade low-speed cascade data by use of the carpet-plotting technique. *NACA T.N.* 3913 (1957).
14. AINLEY, D. G., Performance of axial flow turbines. *Proc. Instn. Mech. Engrs. London*, **159** (1948).
15. AINLEY, D. G. and MATHIESON, G. C. R., A method of performance estimation for axial flow turbines. *A.R.C. R. and M.* 2974 (1951).
16. ZWEIFEL, O., The spacing of turbomachine blading, especially with large angular deflection. *Brown Boveri Rev.* (Dec. 1945).
17. NATIONAL AERONAUTICS AND SPACE ADMINISTRATION, Aerodynamic design of axial-flow compressors. *NASA SP* 36 (1965).
18. DOWDEN, R. ROSEMARY, *Fluid Flow Measurement—a Bibliography*. British Hydromechanics Research Association (1972).
19. LIEBLEIN, S., Loss and stall analysis of compressor cascades. *Trans. Am. Soc Mech. Engrs.* Series D, **81** (1959).
20. LIEBLEIN, S. and ROUDEBUSH, W. H. Theoretical loss relation for low-speed 2D cascade flow. *NACA T.N.* 3662 (1956).

21. SWANN, W. C., A practical method of predicting transonic compressor performance. *Trans. Am. Soc. Mech. Engrs.* Series A, **83** (1961).
22. DUNHAM, J. and CAME, P. Improvements to the Ainley–Mathieson method of turbine performance prediction. *Trans. Am. Soc. Mech. Engrs.*, Series A, **92** (1970).
23. DUNHAM, J., A review of cascade data on secondary losses in turbines. *Journal Mech. Engineering Sci.*, **12** (1970).
24. CAME, P. M., Secondary loss measurements in a cascade of turbine blades. *Proc. Instn. Mech. Engrs. London.* Conference Publication 3 (1973).
25. HORLOCK, J. H., *Axial-flow Turbines*. Butterworths, London (1966).
26. DUNHAM, J., A parametric method of turbine blade profile design. *Am. Soc. Mech. Engrs.* Paper 74-GT-119 (1974).

PROBLEMS

1. Experimental compressor cascade results suggest that the stalling lift coefficient of a cascade blade may be expressed as

$$C_L \left(\frac{c_1}{c_2}\right)^3 = 2 \cdot 2$$

where c_1 and c_2 are the entry and exit velocities. Find the stalling inlet angle for a compressor cascade of space–chord ratio unity if the outlet air angle is 30 deg.

2. Show, for a turbine cascade, using the angle notation of Fig. 3.24, that the lift coefficient is

$$C_L = 2(s/l)(\tan \alpha_1 + \tan \alpha_2)\cos \alpha_m + C_D \tan \alpha_m$$

where $\tan \alpha_m = \frac{1}{2}(\tan \alpha_2 - \tan \alpha_1)$ and $C_D = \text{Drag}/(\frac{1}{2}\rho c_m^2 l)$.

A cascade of turbine nozzle vanes has a blade inlet angle $\alpha_1' = 0$ deg, a blade outlet angle α_2' of 65·5 deg, a chord length l of 45 mm and an axial chord b of 32 mm. The flow entering the blades is to have zero incidence and an estimate of the deviation angle based upon similar cascades is that δ will be about 1·5 deg at low outlet Mach number. If the blade load ratio ψ_T defined by eqn. (3.51) is to be 0·85, estimate a suitable space/chord ratio for the cascade.

Determine the drag and lift coefficients for the cascade given that the profile loss coefficient

$$\lambda = \Delta p_0/(\tfrac{1}{2}\rho c_2^2) = 0 \cdot 035.$$

3. A compressor cascade is to be designed for the following conditions:

Nominal fluid outlet angle	α_2^*	=	30 deg
Cascade camber angle	θ	=	30 deg
Pitch/chord ratio	s/l	=	1·0
Circular arc camberline	a/l	=	0·5

Using Howell's curves and his formula for nominal deviation, determine the nominal incidence, the actual deviation for an incidence of +2·7 deg and the approximate lift coefficient at this incidence.

4. A compressor cascade is built with blades of circular arc camber line, a space/chord ratio of 1.1 and blade angles of 48 and 21 deg at inlet and outlet. Test data taken from the cascade shows that at zero incidence ($i = 0$) the deviation $\delta = 8·2$ deg and the total pressure loss coefficient $\bar{\omega} = \Delta p_0/(\frac{1}{2}\rho c_1^2) = 0·015$. At positive incidence over a limited range ($0 \leqslant i \leqslant 6°$) the variation of both δ and $\bar{\omega}$ for this particular cascade can be represented with sufficient accuracy by linear approximations, viz.

$$\frac{d\delta}{di} = 0·06, \qquad \frac{d\bar{\omega}}{di} = 0·001$$

where i is in degrees.

For a flow incidence of 5·0 deg determine

(i) the flow angles at inlet and outlet;
(ii) the diffuser efficiency of the cascade;
(iii) the static pressure rise of air with a velocity 50 m/s normal to the plane of the cascade.

Assume density of air is 1·2 kg/m³.

5. (a) A cascade of compressor blades is to be designed to give an outlet air angle a_2 of 30 deg for an inlet air angle a_1 of 50 deg measured from the normal to the plane of the cascade. The blades are to have a *parabolic arc* camber line with $a/l = 0·4$ (i.e. the fractional distance along the chord to the point of maximum camber). Determine the space/chord ratio and blade outlet angle if the cascade is to operate at zero incidence and nominal conditions. You may assume the linear approximation for nominal deflection of Howell's cascade correlation:

$$\epsilon^* = (16 - 0·2\,a_2^*)(3 - s/l) \text{ deg}$$

as well as the formula for nominal deviation:

$$\delta^* = \left[0·23\left(\frac{2a}{l}\right)^2 + \frac{a_2^*}{500}\right]\theta\sqrt{\frac{s}{l}} \text{ deg.}$$

(b) The space/chord ratio is now changed to 0·8, but the blade angles remain as they are in part (a) above. Determine the lift coefficient when the incidence of the flow is 2·0 deg. Assume that there is a linear relationship between ϵ/ϵ^* and $(i-i^*)/\epsilon^*$ over a limited region, viz. at $(i-i^*)/\epsilon^* = 0·2$, $\epsilon/\epsilon^* = 1·15$ and at $i = i^*$, $\epsilon/\epsilon^* = 1$. In this region take $C_D = 0·02$.

6. (a) Show that the pressure rise coefficient $C_p = \Delta p/(\frac{1}{2}\rho c_1^2)$ of a compressor cascade is related to the diffuser efficiency η_D and the total pressure loss coefficient ζ by the following expressions:

$$C_p = \eta_D(1 - \sec^2 a_2/\sec^2 a_1) = 1 - (\sec^2 a_2 + \zeta)/\sec^2 a_1$$

where $\eta_D = \Delta p/\{\frac{1}{2}\rho(c_1^2 - c_2^2)\}$

$\zeta = \Delta p_0/(\frac{1}{2}\rho c_x^2)$

$a_1, \quad a_2 =$ flow angles at cascade inlet and outlet.

(b) Determine a suitable *maximum* inlet flow angle of a compressor cascade having a space/chord ratio 0·8 and $a_2 = 30$ deg when the diffusion factor D is to be limited to 0·6. The definition of diffusion factor which should be used is the early Lieblein formula,[20]

$$D = \left(1 - \frac{\cos a_1}{\cos a_2}\right) + \left(\frac{s}{l}\right)\frac{\cos a_1}{2}(\tan a_1 - \tan a_2).$$

(c) The stagnation pressure loss derived from flow measurements on the above cascade is 149 Pa when the inlet velocity c_1 is 100 m/s at an air density ρ of 1·2 kg/m³. Determine the values of

  (i) pressure rise;
  (ii) diffuser efficiency;
  (iii) drag and lift coefficients.

CHAPTER 4

Axial-flow Turbines: Two-dimensional Theory

Power is more certainly retained by wary measures than by daring counsels.
(TACITUS, *Annals*).

INTRODUCTION

The simplest approach to the study of axial-flow turbines (and also axial-flow compressors) is to assume that the flow conditions prevailing at the mean radius fully represent the flow at all other radii. This two-dimensional analysis at the *pitch-line* can provide a reasonable approximation to the actual flow, if the ratio of blade height to mean radius is small. When this ratio is large, however, as in the final stages of a steam turbine or, in the first stages of an axial compressor, a three-dimensional analysis is required. Some important aspects of three-dimensional flows in axial turbomachines are discussed in Chapter 6. Two further assumptions are, that radial velocities are zero, and that the flow is invariant along the circumferential direction (i.e. there are no "blade-to-blade" flow variations).

In this chapter the presentation of the analysis has been devised with compressible flow effects in mind. This approach is then applicable to both steam and gas turbines provided that, in the former case, the steam condition remains wholly within the *vapour* phase (i.e. superheat region). The situation arising when the nominal state point falls below the dryness line on a Mollier chart becomes extremely complicated and beyond the intended scope of this book. The inquiring student is referred to a paper by Ryley[1] on this subject. Much early work concerning flows in steam turbine nozzles and blade rows are reported in the work of Stodola[2] and Kearton[3] and more recently reviewed by Horlock.[6]

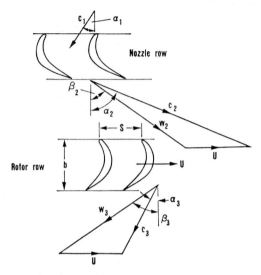

FIG. 4.1. Turbine stage velocity diagram.

VELOCITY DIAGRAM OF THE AXIAL TURBINE STAGE

The axial turbine stage comprises a row of fixed guide vanes or nozzles (often called a *stator* row) and a row of moving blades or buckets (a *rotor* row). Fluid enters the stator with absolute velocity c_1 at angle a_1 and accelerates to an absolute velocity c_2 at angle a_2 (Fig. 4.1). All angles are measured from the axial (x) direction. The *sign convention* is such that angles and velocities as drawn in Fig. 4.1 will be taken as positive throughout this chapter. From the velocity diagram, the rotor inlet *relative* velocity w_2, at an angle β_2, is found by subtracting, vectorially, the blade speed U from the absolute velocity c_2. The relative flow within the rotor accelerates to velocity w_3 at an angle β_3 at rotor outlet; the corresponding absolute flow (c_3, a_3) is obtained by adding, vectorially, the blade speed U to the relative velocity w_3.

The continuity equation for uniform, steady flow is,

$$\rho_1 A_1 c_{x1} = \rho_2 A_2 c_{x2} = \rho_3 A_3 c_{x3}.$$

In two-dimensional theory of turbomachines it is usually assumed, for simplicity, that the axial velocity remains constant i.e. $c_{x1} = c_{x2} = c_{x3} = c_x$.

This must imply that,

$$\rho_1 A_1 = \rho_2 A_2 = \rho_3 A_3 = \text{constant.} \tag{4.1}$$

THERMODYNAMICS OF THE AXIAL TURBINE STAGE

The work done on the rotor by unit mass of fluid, the specific work, equals the stagnation enthalpy drop incurred by the fluid passing through the stage (assuming adiabatic flow), or,

$$\Delta W = \dot{W}/\dot{m} = h_{01} - h_{03} = U(c_{y2} + c_{y3}). \tag{4.2}$$

In eqn. (4.2) the absolute tangential velocity components (c_y) are *added*, so as to adhere to the agreed sign convention of Fig. 4.1. As no work is done in the nozzle row, the stagnation enthalpy across it remains constant and

$$h_{01} = h_{02}. \tag{4.3}$$

Writing $h_0 = h + \frac{1}{2}(c_x^2 + c_y^2)$ and using eqn. (4.3) in eqn. (4.2) we obtain,

$$h_{02} - h_{03} = (h_2 - h_3) + \tfrac{1}{2}(c_{y2}^2 - c_{y3}^2) = U(c_{y2} + c_{y3}),$$

hence,

$$(h_2 - h_3) + \tfrac{1}{2}(c_{y2} + c_{y3})[(c_{y2} - U) - (c_{y3} + U)] = 0.$$

It is observed from the velocity triangles of Fig. 4.1 that $c_{y2} - U = w_{y2}$, $c_{y3} + U = w_{y3}$ and $c_{y2} + c_{y3} = w_{y2} + w_{y3}$. Thus,

$$(h_2 - h_3) + \tfrac{1}{2}(w_{y2}^2 - w_{y3}^2) = 0.$$

Add and subtract $\frac{1}{2}c_x^2$ to the above equation

$$h_2 + \tfrac{1}{2}w_2^2 = h_3 + \tfrac{1}{2}w_3^2 \quad \text{or} \quad h_{02\text{rel}} = h_{03\text{rel}}. \tag{4.4}$$

Thus, we have proved that the *relative* stagnation enthalpy, $h_{0\text{rel}} = h + \frac{1}{2}w^2$, remains unchanged through the rotor of an axial turbomachine. It is implicitly assumed that no radial shift of the streamlines occurs in this flow. In a *radial flow* machine a more general analysis is necessary (see Chapter 7) which takes account of the blade speed change between rotor inlet and outlet.

A Mollier diagram showing the change of state through a complete turbine stage, including the effects of irreversibility, is given in Fig. 4.2.

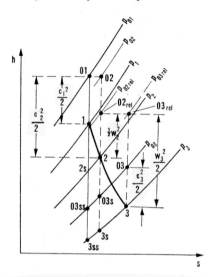

FIG. 4.2. Mollier diagram for a turbine stage.

Through the nozzles, the state point moves from 1 to 2 and the static pressure decreases from p_1 to p_2. In the rotor row, the absolute static pressure reduces (in general) from p_2 to p_3. It is important to note that the conditions contained in eqns. (4.2)–(4.4) are all satisfied in the figure.

STAGE LOSSES AND EFFICIENCY

In Chapter 2 various definitions of efficiency for complete turbomachines were given. For a *turbine stage* the total-to-total efficiency is,

$$\eta_{tt} = \frac{\text{actual work output}}{\text{ideal work output when operating to same back pressure}}$$

$$= (h_{01} - h_{03})/(h_{01} - h_{03ss}).$$

At the entry and exit of a *normal* stage the flow conditions (absolute velocity and flow angle) are identical, i.e. $c_1 = c_3$ and $\alpha_1 = \alpha_3$. If it is assumed that $c_{3ss} = c_3$, which is a reasonable approximation, the total-to-total efficiency becomes

$$\eta_{tt} = (h_1 - h_3)/(h_1 - h_{3ss})$$
$$= (h_1 - h_3)/\{(h_1 - h_3) + (h_3 - h_{3s}) + (h_{3s} - h_{3ss})\}. \tag{4.5}$$

Now the slope of a constant pressure line on a Mollier diagram is $(\partial h/\partial s)_p = T$, obtained from eqn. (2.18). Thus, for a finite change of enthalpy in a constant pressure process, $\Delta h \doteqdot T\Delta s$ and, therefore,

$$h_{3s} - h_{3ss} \doteqdot T_3(s_{3s} - s_{3ss}), \tag{4.6a}$$

$$h_2 - h_{2s} \doteqdot T_2(s_2 - s_{2s}). \tag{4.6b}$$

Noting, from Fig. 4.2, that $s_{3s} - s_{3ss} = s_2 - s_{2s}$, the last two equations can be combined to give

$$h_{3s} - h_{3ss} = (T_3/T_2)(h_2 - h_{2s}). \tag{4.7}$$

The effects of irreversibility through the stator and rotor are expressed by the differences in static enthalpies, $(h_2 - h_{2s})$ and $(h_3 - h_{3s})$ respectively. Non-dimensional enthalpy "loss" coefficients can be defined in terms of the exit kinetic energy from each blade row. Thus, for the nozzle row,

$$h_2 - h_{2s} = \tfrac{1}{2}c_2^2\zeta_N. \tag{4.8a}$$

For the rotor row,

$$h_3 - h_{3s} = \tfrac{1}{2}w_3^2\zeta_R. \tag{4.8b}$$

Combining eqns. (4.7) and (4.8) with eqn. (4.5) gives

$$\eta_{tt} = \left[1 + \frac{\zeta_R w_3^2 + \zeta_N c_2^2 T_3/T_2}{2(h_1 - h_3)}\right]^{-1}. \tag{4.9}$$

When the exit velocity is not recovered (in Chapter 2, examples of such cases are quoted) a total-to-static efficiency for the stage is used.

$$\eta_{ts} = (h_{01} - h_{03})/(h_{01} - h_{3ss})$$
$$= \left[1 + \frac{\zeta_R w_3^2 + \zeta_N c_2^2 T_3/T_2 + c_1^2}{2(h_1 - h_3)}\right]^{-1}, \tag{4.10}$$

where, as before, it is assumed that $c_1 = c_3$.

In initial calculations or, in cases where the static temperature drop through the rotor is not large, the temperature ratio T_3/T_2 is set equal to unity, resulting in the more convenient approximations,

$$\eta_{tt} = \left[1 + \frac{\zeta_R w_3^2 + \zeta_N c_2^2}{2(h_1 - h_3)}\right]^{-1}, \tag{4.9a}$$

$$\eta_{ts} = \left[1 + \frac{\zeta_R w_3^2 + \zeta_N c_2^2 + c_1^2}{2(h_1 - h_3)}\right]^{-1}. \tag{4.10a}$$

So that an assessment can be made of the efficiency of a proposed turbine stage, some means of determining the loss coefficients is required in advance. Several methods of correlation are available but the method of Soderberg,[4] in particular, commends itself for use because of its extreme simplicity and reasonable accuracy.

SODERBERG'S CORRELATION

One method of obtaining design data on turbine blade losses is to assemble information on the overall efficiencies of a wide variety of turbines, and from this calculate the individual blade row losses. This system was developed by Soderberg[4] from a large number of tests performed on steam turbines and on cascades, and extended to fit data obtained from small turbines with very low aspect ratio blading (small height–chord). Soderberg's method was intended only for turbines conforming to the standards of "good design", as discussed below. The method was used by Stenning[5] to whom reference can be made.

A paper by Horlock[6] has critically reviewed several different and widely used methods of obtaining design data for turbines. His paper confirms the claim made for Soderberg's correlation that, although based on relatively few parameters, it is of comparable accuracy with the best of the other methods.

Soderberg found that for the *optimum* space–chord ratio, turbine blade losses (with "full admission" to the complete annulus) could be correlated with space–chord ratio, blade aspect ratio, blade thickness–chord ratio and Reynolds number. Soderberg used *Zweifel's criterion* (see Chapter 3) to obtain the optimum space–chord ratio of turbine cascades based upon the cascade geometry. Zweifel suggested that the aerodynamic load coefficient ψ_T should be approximately 0·8. Following the notation of Fig. 4.1

$$\psi_T = 0.8 = 2(s/b)(\tan \alpha_1 + \tan \alpha_2)\cos^2 \alpha_2. \qquad (4.11)$$

The optimum space–chord ratio may be obtained from eqn. (4.11) for specified values of α_1 and α_2.

For turbine blade rows operating at this load coefficient, with a Reynolds number of 10^5 and aspect ratio H/b = blade height/axial chord) of 3, the "nominal" loss coefficient ζ^* is a simple function of the fluid deflection angle $\varepsilon = \alpha_1 + \alpha_2$, for a given thickness–chord ratio (t_{max}/l). Values of ζ^* are drawn in Fig. 4.3 as a function of deflection ε,

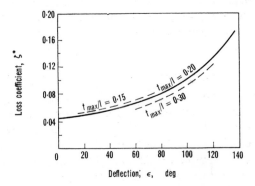

FIG. 4.3. Soderberg's correlation of turbine blade loss coefficient with fluid deflection (adapted from Horlock[6]).

for several ratios of t_{max}/l. A frequently used analytical simplification of this correlation (for $t_{max}/l = 0.2$), which is useful in initial performance calculations, is

$$\zeta^* = 0.04 + 0.06 \left(\frac{\varepsilon}{100}\right)^2. \qquad (4.12)$$

This expression fits Soderberg's curve (for $t_{max}/l = 0.2$) quite well for $\varepsilon \leq 120°$, but is less accurate at higher deflections. For turbine rows operating at zero incidence, which is the basis of Soderberg's correlation, the fluid deflection is little different from the blading deflection since, for *turbine cascades*, deviations are usually small. Thus, for a nozzle row, $\varepsilon = \varepsilon_N = \alpha_2' + \alpha_1'$ and for a rotor row, $\varepsilon = \varepsilon_R = \beta_2' + \beta_3'$ can be used (the prime referring to the actual blade angles).

If the aspect ratio H/b is other than 3, a correction to the nominal loss coefficient ζ^* is made as follows:

for nozzles,

$$1 + \zeta_1 = (1 + \zeta^*)(0.993 + 0.021b/H), \qquad (4.13a)$$

for rotors,

$$1 + \zeta_1 = (1 + \zeta^*)(0.975 + 0.075b/H), \qquad (4.13b)$$

where ζ_1 is the loss coefficient at a Reynolds number of 10^5.

A further correction can be made if the Reynolds number is different from 10^5. As used in this section, Reynolds number is based upon exit velocity c_2 and the hydraulic mean diameter D_h at the throat section.

$$Re = \rho_2 c_2 D_h/\mu, \qquad (4.14)$$

where

$$D_h = 2sH\cos a_2/(s\cos a_2 + H).$$

(N.B. Hydraulic mean diameter $= 4 \times$ flow area $\div$ wetted perimeter.)
The Reynolds number correction is

$$\zeta_2 = \left(\frac{10^5}{Re}\right)^{\frac{1}{4}} \zeta_1. \qquad (4.15)$$

Soderberg's method of loss prediction gives turbine efficiencies with an error of less than 3% over a wide range of Reynolds number and aspect ratio when additional corrections are included to allow for tip leakage and disc friction. An approximate correction for tip clearance may be incorporated by the simple expedient of multiplying the final calculated stage efficiency by the ratio of "blade" area to *total* area (i.e. "blade" area + clearance area).

TYPES OF AXIAL TURBINE DESIGN

The process of choosing the best turbine design for a given application usually involves juggling several parameters which may be of equal importance, for instance, rotor angular velocity, weight, outside diameter, efficiency, so that the final design lies within acceptable limits for each parameter. In consequence, a single presentation can hardly do justice to the real problem. However, a consideration of the factors affecting turbine efficiency for a simplified case can provide a useful guide to the designer.

Consider the problem of selecting an axial turbine design for which the mean blade speed U, the specific work ΔW, and the axial velocity c_x, have already been selected. The upper limit of blade speed is limited by stress; the limit on blade tip speed is roughly 450 m/s although some experimental turbines have been operated at higher speeds. The axial velocity is limited by flow area considerations. It is assumed that the blades are sufficiently short to treat the flow as two-dimensional.

Now

$$\Delta W = U(c_{y2} + c_{y3}).$$

With ΔW, U and c_x fixed the only remaining parameter required to completely define the velocity triangles is c_{y2}, since

$$c_{y3} = \Delta W/U - c_{y2}. \tag{4.16}$$

For different values of c_{y2} the velocity triangles can be constructed, the loss coefficients determined and η_{tt}, η_{ts} calculated. Stenning[10] considered a family of turbines each having a flow coefficient $c_x/U = 0\cdot4$, blade aspect ratio $H/b = 3$ and Reynolds number $Re = 10^5$, and calculated η_{tt}, η_{ts} for stage loading factors $\Delta W/U^2$ of 1, 2 and 3 using Soderberg's correlation. The results of this calculation are shown in Fig. 4.4. It will be noted that these results relate to blading efficiency and make no allowance for losses due to tip clearance and disc friction.

STAGE REACTION

The classification of different types of axial turbine is more conveniently described by the *degree of reaction* or *reaction ratio R*, of each stage rather than by the ratio c_{y2}/U. As a means of description the term reaction has certain inherent advantages which become apparent later. Several definitions of reaction are available; the classical definition is given as the ratio of the static pressure drop in the rotor to the static pressure drop in the stage. However, it is more useful to define the reaction ratio as the static *enthalpy* drop in the rotor to the static

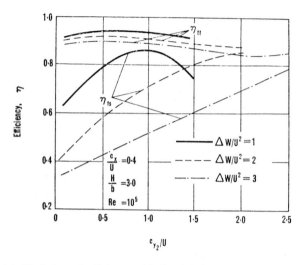

FIG. 4.4. Variation of efficiency with (c_{y2}/U) for several values of stage loading factor $\Delta W/U^2$ (adapted from Shapiro *et al.*[10]).

enthalpy drop in the stage because it then becomes, in effect, a statement of the stage *flow geometry*. Thus,

$$R = (h_2 - h_3)/(h_1 - h_3). \qquad (4.17)$$

If the stage is normal (i.e. $c_1 = c_3$) then,

$$R = (h_2 - h_3)/(h_{01} - h_{03}). \qquad (4.18)$$

Using eqn. (4.4), $h_2 - h_3 = \tfrac{1}{2}(w_3^2 - w_2^2)$ and eqn. (4.18) becomes,

$$R = \frac{w_3^2 - w_2^2}{2U(c_{y2} + c_{y3})}. \qquad (4.19)$$

Assuming constant axial velocity through the stage

$$R = \frac{(w_{y3} - w_{y2})(w_{y3} + w_{y2})}{2U(c_{y2} + c_{y3})} = \frac{w_{y3} - w_{y2}}{2U}, \qquad (4.20)$$

since, upon referring to Fig. 4.1, it is seen that

$$c_{y2} = w_{y2} + U \quad \text{and} \quad c_{y3} = w_{y3} - U. \qquad (4.21)$$

Thus,

$$R = \frac{c_x}{2U} (\tan \beta_3 - \tan \beta_2) \qquad (4.22a)$$

or

$$R = \tfrac{1}{2} + \frac{c_x}{2U} (\tan \beta_3 - \tan \alpha_2), \qquad (4.22b)$$

after using eqn. (4.21).

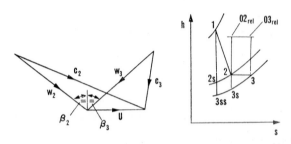

FIG. 4.5. Velocity diagram and Mollier diagram for a zero reaction turbine stage.

If $\beta_3 = \beta_2$, the reaction is zero; if $\beta_3 = \alpha_2$ the reaction is 50%. These two special cases are discussed below in more detail.

Zero reaction stage

From the definition of reaction, when $R = 0$, eqn. (4.18) indicates that $h_2 = h_3$ and eqn. (4.22a) that $\beta_2 = \beta_3$. The Mollier diagram and velocity triangles corresponding to these conditions are sketched in Fig. 4.5. Now as $h_{02rel} = h_{03rel}$ and $h_2 = h_3$ for $R = 0$ it must follow, therefore, that $w_2 = w_3$. It will be observed from Fig. 4.5 that, because of irreversibility, there is a *pressure drop* through the rotor row. The zero reaction stage is *not* the same thing as an *impulse* stage; in the latter case there is, by definition, no pressure drop through the rotor. The Mollier diagram for an impulse stage is shown in Fig. 4.6 where it is seen that the enthalpy *increases* through the rotor. The implication is clear from eqn. (4.18); the reaction is negative for the impulse turbine stage when account is taken of the irreversibility.

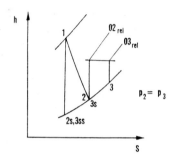

FIG. 4.6. Mollier diagram for an impulse turbine stage.

50 per cent reaction stage

The combined velocity diagram for this case is symmetrical as can be seen from Fig. 4.7, since $\beta_3 = \alpha_2$. Because of the symmetry it is at once obvious that $\beta_2 = \alpha_3$, also. Now with $R = \frac{1}{2}$, eqn. (4.18) implies that the enthalpy drop in the nozzle row equals the enthalpy drop in the rotor, or

$$h_1 - h_2 = h_2 - h_3. \qquad (4.23)$$

Figure 4.7 has been drawn with the same values of c_x, U and ΔW, as in Fig. 4.5 (zero reaction case), to emphasise the difference in flow geometry between the 50% reaction and zero reaction stages.

DIFFUSION WITHIN BLADE ROWS

Any diffusion of the flow through turbine blade rows is particularly undesirable and must, at the design stage, be avoided at all costs. This is because the adverse pressure gradient (arising from the flow diffusion) coupled with large amounts of fluid deflection (usual in turbine blade rows), makes boundary-layer separation more than merely possible, with the result that large scale losses arise. A compressor blade row, on the other hand, is designed to cause the fluid pressure to rise in the direction of flow, i.e. an *adverse* pressure gradient. The magnitude of this gradient is strictly controlled in a compressor, mainly by having a fairly limited amount of fluid deflection in each blade row.

The comparison of the profile losses given in Fig. 3.12 is illustrative of the undesirable result of negative "reaction" in a turbine blade row. The use of the term reaction here needs qualifying as it was only defined with respect to a complete stage. From eqn. (4.22a) the ratio R/ϕ can be expressed for a single row of blades if the flow angles are known. The original data provided with Fig. 3.12 gives the blade inlet angles for impulse and reaction blades as 45·5 and 18·9 deg respectively. Thus, the flow angles can be found from Fig. 3.12 for the range of incidence given, and R/ϕ can be calculated. For the reaction blades R/ϕ decreases

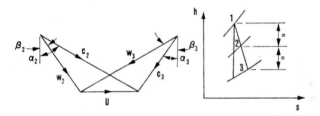

Fig. 4.7. Velocity diagram and Mollier diagram for a 50% reaction turbine stage.

as incidence increases going from 0·36 to 0·25 as i changes from 0 to 10 deg. The impulse blades, which it will be observed have a dramatic increase in blade profile loss, has R/ϕ decreasing from zero to $-0·25$ in the same range of incidence.

It was shown above that negative values of reaction indicated diffusion of the rotor relative velocity (i.e. for $R < 0$, $w_3 < w_2$). A similar condition which holds for diffusion of the nozzle absolute velocity is, that if $R > 1$, $c_2 < c_1$.

Substituting $\tan \beta_3 = \tan \alpha_3 + U/c_x$ into eqn. (4.22b) gives

$$R = 1 + \frac{c_x}{2U} (\tan \alpha_3 - \tan \alpha_2). \qquad (4.22c)$$

Thus, when $\alpha_3 = \alpha_2$, the reaction is unity (also $c_2 = c_3$). The velocity diagram for $R = 1$ is shown in Fig. 4.8 with the same values of c_x, U and ΔW used for $R = 0$ and $R = \frac{1}{2}$. It will be apparent that if R exceeds unity, then $c_2 < c_1$ (i.e. nozzle flow *diffusion*).

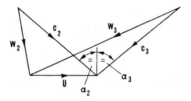

FIG. 4.8. Velocity diagram for 100% reaction turbine stage.

EXAMPLE 1: A single-stage gas turbine operates at its design condition with an axial absolute flow at entry and exit from the stage. The absolute flow angle at nozzle exit is 70 deg. At stage entry the total pressure and temperature are 311 kPa and 850°C respectively. The exhaust static pressure is 100 kPa, the total-to-static efficiency is 0·87 and the mean blade speed is 500 m/s.

Assuming constant axial velocity through the stage, determine

  (i) the specific work done;
 (ii) the Mach number leaving the nozzle;
(iii) the axial velocity;
(iv) the total-to-total efficiency;
 (v) the stage reaction.

Take $C_p = 1·148$ kJ/(kg°C) and $\gamma = 1·33$ for the gas.

Solution: (i) From eqn. (4.10), total-to-static efficiency is

$$\eta_{ts} = \frac{h_{01} - h_{03}}{h_{01} - h_{3ss}} = \frac{\Delta W}{h_{01}\{1 - (p_3/p_{01})^{(\gamma - 1)/\gamma}\}}.$$

Thus, the specific work is

$$\Delta W = \eta_{ts}C_pT_{01}\{1 - (p_3/p_{01})^{(\gamma - 1)/\gamma}\} = 0·87 \times 1148 \times 1123$$
$$\times \{1 - (1/3·11)^{0·248}\}$$
$$= 276 \text{ kJ/kg.}$$

(ii) At nozzle exit the Mach number is

$$M_2 = c_2/(\gamma RT_2)^{\frac{1}{2}}$$

and it is necessary to solve the velocity diagram to find c_2 and hence to determine T_2.

As
$$c_{y3} = 0, \quad \Delta W = U c_{y2}$$

$$c_{y2} = \frac{\Delta W}{U} = \frac{276 \times 10^3}{500} = 552 \text{ m/s}$$

$$c_2 = c_{y2}/\sin \alpha_2 = 588 \text{ m/s}.$$

Referring to Fig. 4.2, across the nozzle $h_{01} = h_{02} = h_2 + \frac{1}{2}c_2^2$, thus

$$T_2 = T_{01} - \frac{1}{2}c_2^2/C_p = 973 \text{ K}.$$

Hence, $M_2 = 0.97$ with $\gamma R = (\gamma - 1)C_p$.

(iii) The axial velocity, $c_x = c_2 \cos \alpha_2 = 200$ m/s.
(iv) $\eta_{tt} = \Delta W/(h_{01} - h_{3ss} - \frac{1}{2}c_3^2)$.

After some rearrangement,

$$\frac{1}{\eta_{tt}} = \frac{1}{\eta_{ts}} - \frac{c_3^2}{2\,\Delta W} = \frac{1}{0.87} - \frac{200^2}{2 \times 276 \times 10^3} = 1.0775.$$

Therefore
$$\eta_{tt} = 0.93.$$

(v) Using eqn. (4.22a), the reaction is

$$R = \frac{1}{2}(c_x/U)(\tan \beta_3 - \tan \beta_2).$$

From the velocity diagram, $\tan \beta_3 = U/c_x$ and $\tan \beta_2 = \tan \alpha_2 - U/c_x$

$$R = 1 - \frac{1}{2}(c_x/U)\tan \alpha_2 = 1 - 200 \times 0.2745/1000$$

$$= 0.451.$$

EXAMPLE 2: Verify the assumed value of total-to-static efficiency in the above example using Soderberg's correlation method. The average blade aspect ratio for the stage $H/b = 5.0$, the maximum blade thickness–chord ratio is 0.2 and the average Reynolds number, defined by eqn. (4.14), is 10^5.

The approximation for total-to-static efficiency, eqn. (4.10a), is used and can be rewritten as

$$\frac{1}{\eta_{ts}} = 1 + \frac{\zeta_R(w_3/U)^2 + \zeta_N(c_2/U)^2 + (c_x/U)^2}{2\,\Delta W/U^2}.$$

The loss coefficients ζ_R and ζ_N, uncorrected for the effects of blade aspect ratio, are determined using eqn. (4.12) which requires a knowledge of flow turning angle ϵ for each blade row.

For the nozzles, $\alpha_1 = 0$ and $\alpha_2 = 70$ deg, thus $\epsilon_N = 70$ deg.

$$\zeta_N{}^* = 0{\cdot}04(1 + 1{\cdot}5 \times 0{\cdot}7^2) = 0{\cdot}0694.$$

Correcting for aspect ratio with eqn. (4.13a),

$$\zeta_{N1} = 1{\cdot}0694(0{\cdot}993 + 0{\cdot}021/5) - 1 = 0{\cdot}0666.$$

For the rotor, $\tan \beta_2 = (c_{y2} - U)/c_x = (552 - 500)/200 = 0{\cdot}26$,

$$\beta_2 = 14{\cdot}55 \text{ deg}.$$

Therefore

$$\tan \beta_3 = U/c_x = 2{\cdot}5,$$

and

$$\beta_3 = 68{\cdot}2 \text{ deg}.$$

Therefore

$$\epsilon_R = \beta_2 + \beta_3 = 82{\cdot}75 \text{ deg},$$

$$\zeta_R{}^* = 0{\cdot}04(1 + 1{\cdot}5 \times 0{\cdot}8275^2) = 0{\cdot}0812.$$

Correcting for aspect ratio with eqn. (4.13b)

$$\zeta_{R1} = 1{\cdot}0812(0{\cdot}975 + 0{\cdot}075/5) - 1 = 0{\cdot}0712.$$

The velocity ratios are:

$$\left(\frac{w_3}{U}\right)^2 = 1 + \left(\frac{c_x}{U}\right)^2 = 1{\cdot}16$$

$$\left(\frac{c_2}{U}\right)^2 = \left(\frac{588}{500}\right)^2 = 1{\cdot}382; \quad \left(\frac{c_x}{U}\right)^2 = 0{\cdot}16$$

and the stage loading factor is,

$$\frac{\Delta W}{U^2} = \frac{c_{y2}}{U} = \frac{552}{500} = 1{\cdot}104$$

Therefore $\dfrac{1}{\eta_{ts}} = 1 + \dfrac{0{\cdot}0712 \times 1{\cdot}16 + 0{\cdot}0666 \times 1{\cdot}382 + 0{\cdot}16}{2 \times 1{\cdot}104}$

$$= 1 + 0{\cdot}1515$$

Thus $\eta_{ts} = 0.869$.

This result is very close to the value assumed in the first example.

It is not too difficult to include the temperature ratio T_3/T_2 implicit in the more exact eqn. (4.10) in order to see how little effect the correction will have. To calculate T_3

$$T_3 = T_{01} - \frac{\Delta W + \frac{1}{2}c_3{}^2}{C_p} = 1123 - \frac{276{,}000 + 20{,}000}{1148}$$

$$= 865 \text{ K}.$$

$T_3/T_2 = 865/973 = 0.89$.

Therefore $\dfrac{1}{\eta_{ts}} = 1 + \dfrac{0.0712 \times 1.16 + 0.89 \times 0.0666 \times 1.382 + 0.16}{2 \times 1.104}$

$$= 1 + 0.1468.$$

Hence $\eta_{ts} = 0.872$.

CHOICE OF REACTION AND EFFECT ON EFFICIENCY

In Fig. 4.4 the total-to-total and total-to-static efficiencies are shown plotted against c_{y2}/U for several values of stage loading factor $\Delta W/U^2$. These curves can now easily be replotted against the degree of reaction R instead of c_{y2}/U. Equation (4.22c) can be rewritten as $R = 1 + (c_{y3} - c_{y2})/(2U)$ and c_{y3} eliminated using eqn. (4.16) to give

$$R = 1 + \frac{\Delta W}{2U^2} - \frac{c_{y2}}{U}. \qquad (4.24)$$

The replotted curves are shown in Fig. 4.9 as presented by Shapiro et al.[10] In the case of total-to-static efficiency, it is at once apparent that this is optimised, at a given blade loading, by a suitable choice of reaction. When $\Delta W/U^2 = 2$, the maximum value of η_{ts} occurs with approximately zero reaction. With lighter blade loading, the optimum η_{ts} is obtained with higher reaction ratios. When $\Delta W/U^2 > 2$, the highest value of η_{ts} attainable without rotor *relative* flow diffusion occurring, is obtained with $R = 0$.

From Fig. 4.4, for a fixed value of $\Delta W/U^2$, there is evidently only a relatively small change in total-to-total efficiency (compared with η_{ts}),

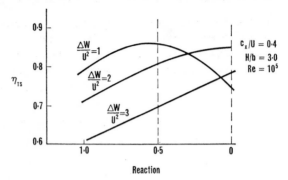

FIG. 4.9. Influence of reaction on total-to-static efficiency with fixed values of stage loading factor.

for a wide range of possible designs. Thus η_{tt} is not greatly affected by the choice of reaction. However, the maximum value of η_{tt} decreases as the stage loading factor increases. To obtain high total-to-total efficiency, it is therefore necessary to use the highest possible value of blade speed consistent with blade stress limitations (i.e. to reduce $\Delta W/U^2$).

MAXIMUM TOTAL-TO-STATIC EFFICIENCY OF A REVERSIBLE TURBINE STAGE

When blade losses and exit kinetic energy loss are included in the definition of efficiency, we have shown, eqn. (4.10a), that the efficiency is

$$\eta_{ts} = \frac{h_{01} - h_{03}}{h_{01} - h_{3ss}} = \left[1 + \frac{w_3^2 \zeta_r + c_2^2 \zeta_n + c_3^2}{2(h_1 - h_3)} \right]^{-1}.$$

In the case of the *ideal* (or reversible) turbine stage the only loss is due to the exhaust kinetic energy and then the total-to-static efficiency is

$$\eta_{ts} = \frac{h_{01} - h_{03ss}}{h_{01} - h_{3ss}} = \left[1 + \frac{c_3^2}{2 U(c_{y2} + c_{y3})} \right]^{-1} \qquad (4.25a)$$

since $\Delta W = h_{01} - h_{03ss} = U(c_{y2} + c_{y3})$ and $h_{03ss} - h_{3ss} = \frac{1}{2}c_3^2$.

The maximum value of η_{ts} is obtained when the exit velocity c_3 is *nearly* a minimum for given turbine stage operating conditions (R, ϕ and a_2). On first thought it may appear obvious that maximum η_{ts} will be obtained when c_3 is absolutely axial (i.e. $a_3 = 0°$) but this is incorrect. By allowing the exit flow to have some counterswirl (i.e. $a_3 > 0$ deg) the work done is increased for only a relatively small increase in the exit kinetic energy loss. Two analyses are now given to show how the total-to-static efficiency of the ideal turbine stage can be optimised for specified conditions.

Substituting $c_{y2} = c_x \tan a_2$, $c_{y3} = c_x \tan a_3$, $c_3 = c_x/\cos a_3$ and $\phi = c_x/U$ into eqn. (4.25), leads to

$$\eta_{ts} = \left[1 + \frac{\phi(1 + \tan^2 a_3)}{2(\tan a_2 + \tan a_3)}\right]^{-1} \qquad (4.25b)$$

i.e.
$$\eta_{ts} = \text{fn} \,(\phi, a_2, a_3).$$

(i) *To find the optimum η_{ts} when R and ϕ are specified*

From eqn. (4.22c) the nozzle flow outlet angle a_2 can be expressed in terms of R, ϕ and a_3 as

$$\tan a_2 = \tan a_3 + 2(1 - R)/\phi. \qquad (4.26)$$

Substituting into eqn. (4.25b)

$$\eta_{ts} = \left[1 + \frac{\phi^2 \,(1 + \tan^2 a_3)}{4 \,(\phi \tan a_3 + 1 - R)}\right]^{-1}.$$

Differentiating this expression with respect to $\tan a_3$, and equating the result to zero,

$$\tan^2 a_3 + 2k \tan a_3 - 1 = 0$$

where $k = (1 - R)/\phi$. This quadratic equation has the solution

$$\tan a_3 = -k + \sqrt{(k^2 + 1)} \qquad (4.27)$$

the value of a_3 being the optimum flow outlet angle from the stage when R and ϕ are specified. From eqn. (4.26), $k = (\tan_2 - \tan a_3)/2$ which when substituted into eqn. (4.27) and simplified gives

$$\tan a_3 = \cot a_2 = \tan(\pi/2 - a_2).$$

Hence, the exact result that

$$a_3 = \pi/2 - a_2.$$

The corresponding idealised $\eta_{ts_{max}}$ and R are

$$\eta_{ts_{max}} = [1 + (\phi/2) \cot a_2]^{-1} \qquad (4.28a)$$

$$R = 1 - \phi(\tan a_2 - \cot a_2)/2.$$

(ii) *To find the optimum η_{ts} when a_2 and ϕ are specified*

Differentiating eqn. (4.25b) with respect to $\tan a_3$ and equating the result to zero,

$$\tan^2 a_3 + 2 \tan a_2 \tan a_3 - 1 = 0.$$

Solving this quadratic, the relevant root is

$$\tan a_3 = \sec a_2 - \tan a_2.$$

Using simple trignometric relations this simplifies still further to

$$a_3 = (\pi/2 - a_2)/2.$$

Substituting this expression for a_3 into eqn. (4.25b) the idealised maximum η_{ts} is obtained

$$\eta_{ts_{max}} = [1 + \phi(\sec a_2 - \tan a_2)]^{-1}. \qquad (4.28b)$$

The corresponding expressions for the degree of reaction R and stage loading coefficient $\Delta W/U^2$ are

$$R = 1 - \phi(\tan a_2 - \tfrac{1}{2} \sec a_2)$$

$$\frac{\Delta W}{U^2} = \phi \sec a_2 = \frac{c_2}{U}. \qquad (4.29)$$

It is interesting that in this analysis the exit swirl angle a_3 is only half that of the constant reaction case. The difference is merely the outcome of the two different sets of constraints used for the two analyses.

For both analyses, as the flow coefficient is reduced towards zero, a_2 approaches $\pi/2$ and a_3 approaches zero. Thus, for such high nozzle exit angle turbine stages, the appropriate blade loading factor for maximum

η_{ts} can be specified if the reaction is known (and conversely). For a turbine stage of 50% reaction (and with $a_3 \to 0$ deg) the appropriate velocity diagram shows that $\Delta W/U^2 \doteq 1$ for maximum η_{ts}. Similarly, a turbine stage of zero reaction (which is an *impulse* stage for ideal, reversible flow) has a blade loading factor $\Delta W/U^2 \doteq 2$ for maximum η_{ts}.

Calculations of turbine stage performance have been made by Horlock[11] both for the reversible and irreversible cases with $R = 0$ and 50%. Figure 4.10 shows the effect of blade losses, determined with Soderberg's correlation, on the total-to-static efficiency of the turbine stage for the constant reaction of 50%. It is evident that exit losses become increasingly dominant as the flow coefficient is increased.

TURBINE FLOW CHARACTERISTICS

An accurate knowledge of the flow characteristics of a turbine is of considerable practical importance as, for instance, in the matching of flows between a compressor and turbine of a jet engine. When a turbine can be expected to operate close to its design incidence (i.e. in the low loss region) the turbine characteristics can be reduced to a single curve. Figure 4.11, due to Mallinson and Lewis,[7] shows a comparison of typical characteristics for one, two and three stages plotted as turbine overall pressure ratio p_{0II}/p_{0I} against a mass flow coefficient $\dot{m}(\sqrt{T_{01}})/p_{0I}$. There is a noticeable tendency for the characteristic to become more ellipsoidal as the number of stages is increased. At a given pressure ratio the mass flow coefficient, or "swallowing capacity" tends to decrease with the addition of further stages to the turbine. One of the earliest attempts to assess the flow variation of a multistage turbine is credited to Stodola,[2] who formulated the much used "ellipse law". The curve labelled "multistage" in Fig. 4.11 is in agreement with the "ellipse law" expression

$$m(\sqrt{T_{01}})/p_{0I} = k[1 - (p_{0II}/p_{0I})^2]^{\frac{1}{2}}, \qquad (4.30)$$

where k is a constant.

This expression has been used for many years in steam turbine practice, but an accurate estimate of the variation in swallowing capacity with pressure ratio is of even greater importance in gas turbine

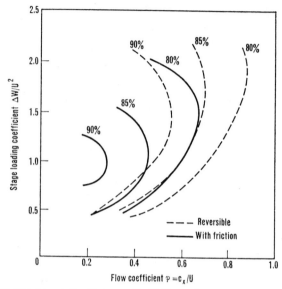

Fɪɢ. 4.10. Total-to-static efficiency of a 50% reaction axial flow turbine stage (adapted from Horlock[11]).

technology. Whereas the average condensing steam turbine, even at part-load, operates at very high pressure ratios, some gas turbines may work at rather low pressure ratios, making flow matching with a compressor a difficult problem. The constant value of swallowing capacity, reached by the single-stage turbine at a pressure ratio a little above 2, and the other turbines at progressively higher pressure ratios, is associated with choking (sonic) conditions in the turbine stator blades.

FLOW CHARACTERISTIC OF A MULTISTAGE TURBINE

Several derivations of the ellipse law are available in the literature. The derivation given below is a slightly amplified version of the proof given by Horlock.[8] A more general method has been given by Egli[9] which takes into consideration the effects when operating outside the normal low loss region of the blade rows.

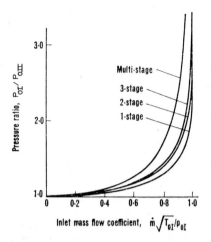

FIG. 4.11. Turbine flow characteristics (after Mallinson and Lewis[7]).

Consider a turbine comprising a large number of normal stages, each of 50% reaction; then, referring to the velocity diagram of Fig. 4.12a, $c_1 = c_3 = w_2$ and $c_2 = w_3$. If the blade speed is maintained constant and the mass flow is reduced, the fluid angles at exit from the rotor (β_3) and nozzles (α_2) will remain constant and the velocity diagram then assumes the form shown in Fig. 4.12b. The turbine, if operated in this manner, will be of low efficiency, as the fluid direction at inlet to each blade row is likely to produce a negative incidence stall. To maintain high efficiency the fluid inlet angles must remain fairly close to the design values. It is therefore assumed that the turbine operates at its highest efficiency at *all off-design conditions* and, by implication, the blade speed is changed in direct proportion to the axial velocity. The velocity triangles are similar at off-design flows but of different scale.

Now the work done by unit mass of fluid through one stage is $U(c_{y2}+c_{y3})$ so that, assuming a perfect gas,

$$C_p \Delta T_0 = C_p \Delta T = U c_x (\tan \alpha_2 + \tan \alpha_3)$$

and, therefore,

$$\Delta T \propto c_x^2.$$

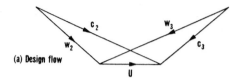

(a) Design flow

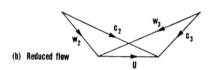

(b) Reduced flow

FIG. 4.12. Change in turbine stage velocity diagram with mass flow at constant blade speed.

Denoting design conditions by subscript d, then

$$\frac{\Delta T}{\Delta T_d} = \left(\frac{c_x}{c_{xd}}\right)^2 \tag{4.31}$$

for equal values of c_x/U.

From the continuity equation, at off-design, $\dot{m} = \rho A c_x = \rho_1 A_1 c_{x1}$, and at design, $\dot{m}_d = \rho_d A c_{xd} = \rho_1 A_1 c_{x1d}$, hence

$$\frac{c_x}{c_{xd}} = \frac{\rho_d}{\rho}\frac{c_{x1}}{c_{x1d}} = \frac{\rho_d}{\rho}\frac{\dot{m}}{\dot{m}_d}. \tag{4.32}$$

Consistent with the assumed mode of turbine operation, the polytropic efficiency is taken to be constant at off-design conditions and, from eqn. (2.37), the relationship between temperature and pressure is therefore,

$$T/p^{\eta_p(\gamma-1)/\gamma} = \text{constant.}$$

Combined with $p/\rho = RT$ the above expression gives, on eliminating p, $\rho/T^n = $ constant, hence

$$\frac{\rho}{\rho_d} = \left(\frac{T}{T_d}\right)^n, \tag{4.33}$$

where $n = \gamma/\{\eta_p(\gamma-1)\}-1$.

For an infinitesimal temperature drop eqn. (4.31) combined with eqns. (4.32) and 4.(33) gives, with little error,

$$\frac{\mathrm{d}T}{\mathrm{d}T_d} = \left(\frac{c_x}{c_{xd}}\right)^2 = \left(\frac{T_d}{T}\right)^{2n}\left(\frac{\dot{m}}{\dot{m}_d}\right)^2. \tag{4.34}$$

Integrating eqn. (4.34),

$$T^{2n+1} = \left(\frac{\dot{m}}{\dot{m}_d}\right)^2 T_d^{2n+1} + K,$$

where K is an arbitrary constant.

To establish a value for K it is noted that if the turbine entry temperature is constant $T_d = T_1$ and $T = T_1$ also.

Thus, $K = [1-(\dot{m}/\dot{m}_d)^2]T_1^{2n+1}$ and

$$\left(\frac{T}{T_1}\right)^{2n+1} - 1 = \left(\frac{\dot{m}}{\dot{m}_d}\right)^2\left[\left(\frac{T_d}{T_1}\right)^{2n+1} - 1\right]. \tag{4.35}$$

Equation (4.35) can be rewritten in terms of pressure ratio since $T/T_1 = (p/p_1)^{\eta_p(\gamma-1)/\gamma}$. As $2n + 1 = 2\gamma/[\eta_p(\gamma - 1)] - 1$ then,

$$\frac{\dot{m}}{\dot{m}_d} = \left\{\frac{1 - (p/p_1)^{2 - \eta_p(\gamma-1)/\gamma}}{1 - (p_d/p_1)^{2 - \eta_p(\gamma-1)/\gamma}}\right\}^{\frac{1}{2}} \tag{4.36a}$$

With $\eta_p = 0.9$ and $\gamma = 1.3$ the pressure ratio index is about 1·8; thus the approximation is often used

$$\frac{\dot{m}}{\dot{m}_d} = \left\{\frac{1 - (p/p_1)^2}{1 - (p_d/p_1)^2}\right\}^{\frac{1}{2}}, \tag{4.36b}$$

which is the ellipse law of a multistage turbine.

REFERENCES

1. RYLEY, D. J., Phase equilibrium in low-pressure steam turbines. *Int. J. Mech. Sci.* (Aug. 1960).
2. STODOLA, A., *Steam and Gas Turbines*, 6th ed. Peter Smith, New York (1945).
3. KEARTON, W. J., *Steam Turbine Theory and Practice*. 7th ed. Pitman, London (1958).
4. SODERBERG, C. R., Unpublished note. Gas Turbine Laboratory, Massachusetts Institute of Technology (1949).
5. STENNING, A. H., Design of turbines for high energy fuel, low power output applications. D.A.C.L. Report 79, Massachusetts Institute of Technology (1953).

6. HORLOCK, J. H., Losses and efficiencies in axial-flow turbines. *Int. J. Mech. Sci.* **2** (1960).
7. MALLINSON, D. H. and LEWIS, W. G. E., The part-load performance of various gas-turbine engine schemes. *Proc. Instn. Mech. Engrs. London,* **159** (1948).
8. HORLOCK, J. H., A rapid method for calculating the "off-design" performance of compressors and turbines. *Aeronaut. Quart* **9** (1958).
9. EGLI, A., The flow characteristics of variable-speed reaction steam turbines. *Trans. Am. Soc. Mech. Engrs.* **58** (1936).
10. SHAPIRO, A. H., SODERBERG, C. R., STENNING, A. H., TAYLOR, E. S. and HORLOCK, J. H., Notes on Turbomachinery. Department of Mechanical Engineering, Massachusetts Institute of Technology. Unpublished (1957).
11. HORLOCK, J. H. *Axial Flow Turbines.* Butterworths, London (1966).

PROBLEMS

1. Show, for an axial flow turbine stage, that the *relative* stagnation enthalpy across the rotor row does not change. Draw an enthalpy–entropy diagram for the stage labelling all salient points.

Stage reaction for a turbine is defined as the ratio of the static enthalpy drop in the rotor to that in the stage. Derive expressions for the reaction in terms of the flow angles and draw velocity triangles for reactions of zero, 0·5 and 1·0.

2. In a Parsons' reaction turbine the rotor blades are similar to the stator blades but with the angles measured in the opposite direction. The efflux angle relative to each row of blades is 70 deg from the axial direction, the exit velocity of steam from the stator blades is 160 m/s, the blade speed is 152·5 m/s and the axial velocity is constant. Determine the specific work done by the steam per stage.

A turbine of 80% internal efficiency consists of ten such stages as described above and receives steam from the stop valve at 1·5 MPa and 300°C. Determine, with the aid of a Mollier chart, the condition of the steam at outlet from the last stage.

3. Values of pressure (kPa) measured at various stations of a zero-reaction gas turbine stage, all at the mean blade height, are shown in the table given below.

Stagnation pressure		Static pressure	
Nozzle entry	414	Nozzle exit	207
Nozzle exit	400	Rotor exit	200

The mean blade speed is 291 m/s, inlet stagnation temperature 1100 K, and the flow angle at nozzle exit is 70 deg measured from the axial direction. Assuming the magnitude and direction of the velocities at entry and exit of the stage are the same, determine the total-to-total efficiency of the stage. Assume a perfect gas with $C_p = 1·148$ kJ/(kg°C) and $\gamma = 1·333$.

4. In a certain axial flow turbine stage the axial velocity c_x is constant. The absolute velocities entering and leaving the stage are in the axial direction. If the flow coefficient c_x/U is 0·6 and the gas leaves the stator blades at 68·2 deg from the axial direction, calculate:

  (i) the stage loading factor, $\Delta W/U^2$;
  (ii) the flow angles relative to the rotor blades;
  (iii) the degree of reaction;
  (iv) the total-to-total and total-to-static efficiencies.

The Soderberg loss correlation, eqn. (4.12) should be used.

5. A gas turbine stage develops 3·36 MW for a mass flow rate of 27·2 kg/s. The *stagnation* pressure and *stagnation* temperature at stage entry are 772 kPa and 1000K. The axial velocity is constant throughout the stage, the gases entering and leaving the stage without any absolute swirl. At nozzle exit the static pressure is 482 kPa and the flow direction is at 18 deg to the *plane of the wheel*. Determine the axial velocity and degree of reaction for the stage given that the entropy increase in the nozzles is 12·9 J/(kg °C).

Assume that the specific heat at constant pressure of the gas is 1·148 kJ/(kg °C) and the gas constant is 0·287 kJ/(kg °C).

Determine also the total-to-total efficiency of the stage given that the increase in entropy of the gas across the rotor is 2·7 J/(kg °C).

6. Derive an approximate expression for the total-to-total efficiency of a turbine stage in terms of the enthalpy loss coefficients for the stator and rotor when the absolute velocities at inlet and outlet are **not** equal.

A steam turbine stage of high hub/tip ratio is to receive steam at a stagnation pressure and temperature of 1·5 MPa and 325°C respectively. It is designed for a blade speed of 200 m/s and the following *blade* geometry was selected:

	Nozzles	Rotor
Inlet angle, deg	0	48
Outlet angle, deg	70·0	56·25
Space/chord ratio, s/l	0·42	—
Blade length/axial chord ratio, H/b	2·0	2·1
Max. thickness/axial chord	0·2	0·2

The deviation angle of the flow from the rotor row is known to be 3 deg on the evidence of cascade tests at the design condition. In the absence of cascade data for the nozzle row, the designer estimated the deviation angle from the approximation 0·19 θ s/l where θ is the blade camber in degrees. Assuming the incidence onto the nozzles is zero, the incidence onto the rotor 1·04 deg and the axial velocity across the stage is constant, determine:

  (i) the axial velocity;
  (ii) the stage reaction and loading factor;
  (iii) the approximate total-to-total stage efficiency on the basis of Soderberg's loss correlation, assuming Reynolds number effects can be ignored;
  (iv) by means of a large steam chart (Mollier diagram) the stagnation temperature and pressure at stage exit.

CHAPTER 5

Axial-flow Compressors, Pumps and Fans: Two-dimensional Analysis

A solemn, strange and mingled air, 't was sad by fits, by starts was wild.
(W. COLLINS, *The Passions.*)

INTRODUCTION

The idea of using a form of *reversed turbine* as an axial compressor is as old as the reaction turbine itself. It is recorded by Stoney[1] that Sir Charles Parsons obtained a patent for such an arrangement as early as 1884. However, simply reversing a turbine for use as a compressor gives efficiencies which are, according to Howell,[2] less than 40% for machines of high pressure ratio. Parsons actually built a number of these machines (*circa* 1900), with blading based upon improved propeller sections. The machines were used for blast furnace work, operating with delivery pressures between 10 and 100 kPa. The efficiency attained by these early, low pressure compressors was about 55%; the reason for this low efficiency is now attributed to blade stall. A high pressure ratio compressor (550 kPa delivery pressure) was also built by Parsons but is reported by Stoney to have "run into difficulties". The design, comprising two axial compressors in series, was abandoned after many trials, the flow having proved to be unstable (presumably due to *compressor surge*). As a result of low efficiency, axial compressors were generally abandoned in favour of multistage centrifugal compressors with their higher efficiency of 70–80%.

It was not until 1926 that any further development on axial compressors was undertaken when A. A. Griffith outlined the basic principles of his aerofoil theory of compressor and turbine design. The subsequent history of the axial compressor is closely linked with that of the aircraft gas turbine and has been recorded by Cox[3] and Constant.[4] The work of the team under Griffith at the Royal Aircraft Establishment, Farnborough, led to the conclusion (confirmed later by rig tests) that efficiencies of at least 90% could be achieved for 'small' stages, i.e. low pressure ratio stages.

The early difficulties associated with the development of axial-flow compressors stemmed mainly from the fundamentally different nature of the flow process compared with that in axial-flow turbines. Whereas in the axial turbine the flow relative to each blade row is *accelerated*, in axial compressors it is *decelerated*. It is now widely known that although a fluid can be rapidly accelerated through a passage and sustain a small or moderate loss in total pressure the same is not true for a rapid deceleration. In the later case large losses would arise as a result of severe stall caused by a large adverse pressure gradient. So as to limit the total pressure losses during flow diffusion it is necessary for the rate of deceleration (and turning) in the blade passages to be severely restricted. (Details of these restrictions are outlined in Chapter 3 in connection with the correlations of Lieblein and Howell.) It is mainly because of these restrictions that axial compressors need to have many stages for a given pressure ratio compared with an axial turbine which needs only a few. Thus, the reversed turbine experiment tried by Parsons was doomed to a low operating efficiency.

Today, although axial compressors are reported with efficiencies of up to 90% at pressure ratios of 6 or 7 to 1, it seems remarkable to reflect that after over 40 years of continuous development, the machine is still able to pose to both research engineers and designers an abundance of unsolved challenging problems. The belief is widely held that the full potential in efficiency and pressure ratio has yet to be reached, and only from a detailed understanding of the more complicated flow phenomena can this objective be reached. A review of many detailed advanced studies in this field is given by Horlock.[5] More recently a review was made by Gostelow *et al.*[14] concerning the developments in aerodynamic design of subsonic and transonic axial flow compressors. Transonic compressors are a fairly recent development and follow the unexpected discovery of good performance at relative Mach numbers near unity.[14] Compressors of many current jet engines have at least one transonic stage.

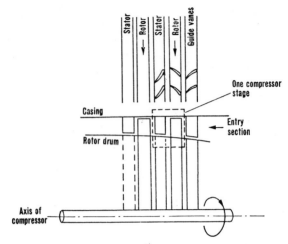

FIG. 5.1. Section and development of the first two stages of a multistage axial compressor.

TWO-DIMENSIONAL ANALYSIS OF THE COMPRESSOR STAGE

The analysis in this chapter is simplified (as it was for axial turbines) by assuming the flow is two-dimensional. This approach can be justified if the blade height is small compared with the mean radius. Again, as for axial turbines, the flow is assumed to be invariant in the circumferential direction and that no spanwise (radial) velocities occur. Some of the three-dimensional effects of axial turbomachines are considered in Chapter 6.

A simplified part section of an axial compressor is shown in Fig. 5.1 together with a projection of the blade rows opened out into a plane array of two-dimensional cascades. A *compressor stage* is defined as a rotor row followed by a stator row, as in the figure. Multistage axial compressors may have as many as twenty stages in some applications, making the machine rather lengthy. The rotor blades are fixed to the *rotor drum* and the stator blades to the *casing*. The inlet guide vanes are *not* regarded as part of the first compressor stage and are treated separately. Their function is different from a stator row since, by directing the flow *away* from the axial direction, they act as a set of turbine nozzles which *accelerate* the flow.

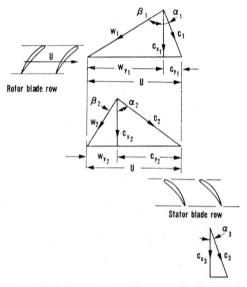

Fig. 5.2. Velocity diagrams for a compressor stage.

VELOCITY DIAGRAMS OF THE COMPRESSOR STAGE

The velocity diagrams for the stage are given in Fig. 5.2 and the convention is adopted throughout this chapter of accepting all angles and swirl velocities in this figure as positive. As for axial turbine stages, a *normal* compressor stage is one where the absolute velocities and flow directions at stage outlet are the same as at stage inlet. The flow from a previous stage (or from the guide vanes) has a velocity c_1 and direction α_1; subtracting vectorially the blade speed U gives the inlet relative velocity w_1 at angle β_1 (the axial direction is the datum for all angles). Relative to the blades of the rotor, the flow is turned to the direction β_2 at outlet with a relative velocity w_2. Clearly, by adding vectorially the blade speed U on to w_2 gives the absolute velocity from the rotor, c_2 at angle α_2. The stator blades deflect the flow towards the axis and the exit velocity is c_3 at angle α_3. For the normal stage $c_3 = c_1$ and $\alpha_3 = \alpha_1$. It will be noticed that as drawn in Fig. 5.2, both the relative velocity in the rotor and the absolute velocity in the stator are diffused. It will be shown later in this chapter, that the relative amount of diffusion of

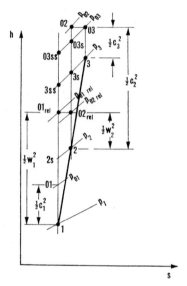

Fig. 5.3. Mollier diagram for an axial compressor stage.

kinetic energy in the rotor and stator rows, significantly influences the stage efficiency.

THERMODYNAMICS OF THE COMPRESSOR STAGE

The specific work done by the rotor on the fluid, from the steady flow energy equation (assuming adiabatic flow) and momentum equation is

$$\Delta W = \dot{W}_p/\dot{m} = h_{02} - h_{01} = U(c_{y2} - c_{y1}). \qquad (5.1)$$

In Chapter 4 it was proved for all axial turbomachines that $h_{0\,\text{rel}}$ ($=h + \frac{1}{2}w^2$) is constant in the rotor. Thus,

$$h_1 + \tfrac{1}{2}w_1^2 = h_2 + \tfrac{1}{2}w_2^2. \qquad (5.2)$$

This is a valid result as long as there is no radial shift of the streamlines across the rotor (i.e. $U_1 = U_2$).

Across the stator, h_0 is constant, and

$$h_2 + \tfrac{1}{2}c_2^2 = h_3 + \tfrac{1}{2}c_3^2. \qquad (5.3)$$

The compression process for the complete stage is represented on a Mollier diagram in Fig. 5.3, which is generalised to include the effects of irreversibility.

STAGE LOSS RELATIONSHIPS AND EFFICIENCY

From eqns. (5.1) and (5.3) the actual work performed by the rotor on unit mass of fluid is $\Delta W = h_{03} - h_{01}$. The reversible or *minimum* work required to attain the same final stagnation pressure as the real process is,

$$\Delta W_{min} = h_{03ss} - h_{01}$$
$$= (h_{03} - h_{01}) - (h_{03s} - h_{03ss}) - (h_{03} - h_{03s})$$
$$\doteqdot \Delta W - T_{03}/T_2(h_2 - h_{2s}) - T_{03}/T_3(h_3 - h_{3s}),$$

using the approximation that $\Delta h = T \Delta s$.

The temperature rise in a compressor stage is only a small fraction of the absolute temperature level and therefore, to a *close* approximation.

$$\Delta W_{min} = \Delta W - (h_2 - h_{2s}) - (h_3 - h_{3s}). \tag{5.4}$$

Again, because of the small stage temperature rise, the density change is also small and it is reasonable to assume incompressibility for the fluid. This approximation is applied *only* to the stage and a *mean* stage density is implied; across a multistage compressor an appreciable density change can be expected.

The enthalpy losses in eqn. (5.4) can be expressed as stagnation pressure losses as follows. As $h_{02} = h_{03}$ then,

$$h_3 - h_2 = \tfrac{1}{2}(c_2^2 - c_3^2)$$
$$= [(p_{02} - p_2) - (p_{03} - p_3)]/\rho, \tag{5.5}$$

since $p_0 - p = \tfrac{1}{2}\rho c^2$ for an incompressible fluid.

Along the isentrope $2-3_s$ in Fig. 5.3, $Tds = 0 = dh - (1/\rho)dp$, and so,

$$h_{3s} - h_2 = (p_3 - p_2)/\rho. \tag{5.6}$$

Thus, subtracting eqn. (5.6) from eqn. (5.5)

$$h_3 - h_{3s} = (p_{02} - p_{03})/\rho = (1/\rho)\Delta p_{0stator}. \tag{5.7}$$

Similarly, for the rotor,

$$h_2 - h_{2s} = (p_{01rel} - p_{02rel})/\rho = (1/\rho)\Delta p_{0rotor}. \tag{5.8}$$

The total-to-total stage efficiency is,

$$\eta_{tt} = \frac{\dot{W}_{pmin}}{\dot{W}_p} \doteq 1 - \frac{(h_2 - h_{2s}) + (h_3 - h_{3s})}{(h_{03} - h_{01})}$$

$$\doteq 1 - \frac{\Delta p_{0stator} + \Delta p_{0rotor}}{\rho(h_{03} - h_{01})}. \tag{5.9}$$

It is to be observed that eqn. (5.9) also has direct application to pumps and fans.

REACTION RATIO

For the case of incompressible and reversible flow it is permissible to define the reaction R, as the ratio of static pressure rise in the rotor to the static pressure rise in the stage

$$R = (p_2 - p_1)/(p_3 - p_1). \tag{5.10a}$$

If the flow is both compressible and irreversible a more general definition of R is the ratio of the rotor static enthalpy rise to the stage static enthalpy rise,

$$R = (h_2 - h_1)/(h_3 - h_1). \tag{5.10b}$$

From eqn. (5.2), $h_2 - h_1 = \frac{1}{2}(w_1^2 - w_2^2)$. For normal stages ($c_1 = c_3$), $h_3 - h_1 = h_{03} - h_{01} = U(c_{y2} - c_{y1})$. Substituting into eqn. (5.10b)

$$R = \frac{w_1^2 - w_2^2}{2U(c_{y2} - c_{y1})} \tag{5.10c}$$

$$= \frac{(w_{y1} + w_{y2})(w_{y1} - w_{y2})}{2U(c_{y2} - c_{y1})},$$

where it is assumed that c_x is constant across the stage. From Fig. 5.2, $c_{y2} = U - w_{y2}$ and $c_{y1} = U - w_{y1}$ so that $c_{y2} - c_{y1} = w_{y1} - w_{y2}$. Thus,

$$R = (w_{y1} + w_{y2})/(2U) = (c_x/U)\tan \beta_m, \tag{5.11}$$

where

$$\tan \beta_m = \frac{1}{2}(\tan \beta_1 + \tan \beta_2). \tag{5.12}$$

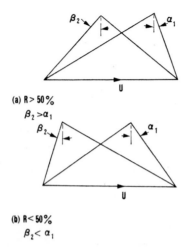

(a) R> 50%
$\beta_2 > \alpha_1$

(b) R< 50%
$\beta_2 < \alpha_1$

FIG. 5.4. Asymmetry of velocity diagrams for reactions greater or less than 50%.

An alternative useful expression for reaction can be found in terms of the fluid outlet angles from each blade row in a stage. With $w_{y1} = U - c_{y1}$, eqn. (5.11) gives,

$$R = \tfrac{1}{2} + (\tan \beta_2 - \tan \alpha_1)c_x/(2U). \tag{5.13}$$

Both expressions for reaction given above may be derived on a basis of incompressible, reversible flow, together with the definition of reaction in eqn. (5.10a).

CHOICE OF REACTION

The reaction ratio is a design parameter which has an important influence on stage efficiency. Stages having 50% reaction are widely used as the adverse (retarding) pressure gradient through the rotor and stator rows is equally shared. This choice of reaction minimises the tendency of the blade boundary layers to separate from the solid surfaces, thus avoiding large stagnation pressure losses.

If $R = 0.5$, then $\alpha_1 = \beta_2$ from eqn. (5.13), and the velocity diagram is symmetrical. The stage enthalpy rise is equally distributed between the rotor and stator rows.

If $R > 0.5$ then $\beta_2 > \alpha_1$ and the velocity diagram is skewed to the *right* as shown in Fig. 5.4a. The static enthalpy rise in the rotor exceeds that in the stator (this is also true for the static pressure rise).

If $R < 0.5$ then $\beta_2 < \alpha_1$ and the velocity diagram is skewed to the *left* as indicated in Fig. 5.4b. Clearly, the stator enthalpy (and pressure) rise exceeds that in the rotor.

In axial turbines the limitation on stage work output is imposed by rotor blade stresses but, in axial compressors, stage performance is limited by Mach number considerations. If Mach number effects could be ignored, the permissible temperature rise, based on incompressible flow cascade limits, increases with the amount of reaction. With a limit of 0·7 on the allowable Mach number, the temperature rise and efficiency are at a maximum with a reaction of 50%.[5]

STAGE LOADING

The stage loading factor ψ is another important design parameter of a compressor stage and is one which strongly affects the off-design performance characteristics. It is defined by

$$\psi = \frac{h_{03} - h_{01}}{U^2} = \frac{c_{y2} - c_{y1}}{U}. \tag{5.14a}$$

With $c_{y2} = U - w_{y2}$ this becomes,

$$\psi = 1 - \phi \, (\tan \alpha_1 + \tan \beta_2), \tag{5.14b}$$

where $\phi = c_x/U$ is called the *flow coefficient*.

The stage loading factor may also be expressed in terms of the lift and drag coefficients for the *rotor*. From Fig. 3.5, replacing α_m with β_m, the tangential blade force on the *moving* blades per unit span is,

$$Y = L \cos \beta_m + D \sin \beta_m$$

$$= L \cos \beta_m \left(1 + \frac{C_D}{C_L} \tan \beta_m\right),$$

where $\tan \beta_m = \frac{1}{2}(\tan \beta_1 + \tan \beta_2)$.

Now $C_L = L/(\frac{1}{2}\rho w_m^2 l)$ hence substituting for L above,

$$Y = \frac{1}{2}\rho c_x^2 l C_L \sec \beta_m (1 + \tan \beta_m C_D/C_L). \tag{5.15}$$

The work done by *each* moving blade per second is YU and is transferred to the fluid flowing through *one* blade passage during that period. Thus, $YU = \rho s c_x (h_{03} - h_{01})$.

Therefore, the stage loading factor may now be written

$$\psi = \frac{h_{03} - h_{01}}{U^2} = \frac{Y}{\rho s c_x U}. \tag{5.16}$$

Substituting eqn. (5.15) in eqn. (5.16) the final result is

$$\psi = (\phi/2) \sec \beta_m (l/s)(C_L + C_D \tan \beta_m). \tag{5.17}$$

In Chapter 3, the approximate analysis indicated that maximum efficiency is obtained when the mean flow angle is 45 deg. The corresponding optimum stage loading factor at $\beta_m = 45$ deg is,

$$\psi_{opt} = (\phi/\sqrt{2})(l/s)(C_L + C_D). \tag{5.18}$$

Since $C_D \ll C_L$ in the normal low loss operating range, it is permissible to drop C_D from eqn. (5.18).

SIMPLIFIED OFF-DESIGN PERFORMANCE

Horlock[5] has considered how the stage loading behaves with varying flow coefficient, ϕ and how this off-design performance is influenced by the choice of design conditions. Now cascade data suggests that fluid *outlet angles* β_2 (for the rotor) and $\alpha_1 (= \alpha_3)$ for the stator, *do not change appreciably* for a range of incidence up to the stall point. The simplification may therefore be made that, for a given stage,

$$\tan \alpha_1 + \tan \beta_2 = t = \text{constant}. \tag{5.19}$$

Inserting this expression into eqn. (5.14b) gives

$$\psi = 1 - \phi t. \tag{5.20a}$$

An inspection of eqns. (5.20a) and (5.14a) indicates that the stagnation enthalpy rise of the stage increases as the mass flow is reduced, when running at constant rotational speed, provided t is positive. The effect is shown in Fig. 5.5, where ψ is plotted against ϕ for several values of t.

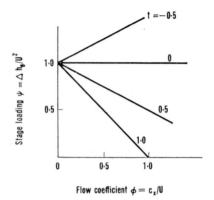

Writing $\psi = \psi_d$ and $\phi = \phi_d$ for conditions at the design point, then

$$\psi_d = 1 - \phi_d t. \tag{5.20b}$$

The values of ψ_d and ϕ_d chosen for a particular stage design, determines the value of t. Thus t is fixed without regard to the degree of reaction and, therefore, the variation of stage loading at off-design conditions is not dependent on the choice of design reaction. However, from eqn. (5.13) it is apparent that, except for the case of 50% reaction when $\alpha_1 = \beta_2$, the reaction *does* change away from the design point. For design reactions exceeding 50% $(\beta_2 > \alpha_1)$, the reaction decreases towards 50% as ϕ decreases; conversely, for design reactions less than 50% the reaction approaches 50% with diminishing flow coefficient.

If t is eliminated between eqns. (5.20a) and (5.20b) the following expression results,

$$\frac{\psi}{\psi_d} = \frac{1}{\psi_d} - \frac{\phi}{\phi_d}\left(\frac{1 - \psi_d}{\psi_d}\right). \tag{5.21}$$

This equation shows that, for a given design stage loading ψ_d, the fractional change in stage loading corresponding to a fractional change in flow coefficient is always the same, independent of the stage reaction. In Fig. 5.6 it is seen that heavily loaded stages ($\psi_d \rightarrow 1$) are the most.

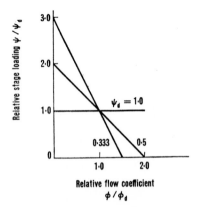

FIG. 5.6. Effect of design stage loading (ψd) on simplified off-design performance characteristics (adapted from Horlock[5]).

flexible producing little variation of ψ with change of ϕ. Lightly loaded
stages ($\psi_d \to 0$) produce large changes in ψ with changing ϕ. Data from
cascade tests show that ψ_d is limited to the range 0·3 to 0·4 for the most
efficient operation and so substantial variations of ψ can be expected
away from the design point.

In order to calculate the pressure rise at off-design conditions the
variation of stage efficiency with flow coefficient is required. For an
ideal stage (no losses) the pressure rise in incompressible flow is given
by

$$\psi = \frac{\Delta h}{U^2} = \frac{\Delta p}{\rho U^2}. \tag{5.22}$$

STAGE PRESSURE RISE

Consider first the *ideal* compressor stage which has no stagnation
pressure losses. Across the rotor row $p_{0\text{rel}}$ is constant and so

$$p_2 - p_1 = \tfrac{1}{2}\rho(w_1^2 - w_2^2). \tag{5.23a}$$

Across the stator row p_0 is constant and so

$$p_3 - p_2 = \tfrac{1}{2}\rho(c_2^2 - c_3^2). \tag{5.23b}$$

Adding together the pressure rise for each row and considering a normal stage ($c_3 = c_1$), gives

$$(p_3 - p_1)2/\rho = (c_2^2 - w_2^2) + (w_1^2 - c_1^2). \tag{5.24}$$

For either velocity triangle (Fig. 5.2), the Cosine Rule gives $c^2 - U^2 + w^2 = 2Uw \cos(\pi/2 - \beta)$ or

$$c^2 - w^2 = U^2 - 2Uw_y. \tag{5.25}$$

Substituting eqn. (5.25) into the stage pressure rise,

$$2(p_3 - p_1)/\rho = (U^2 - 2Uw_{y2}) - (U^2 - 2Uw_{y1})$$

$$= 2U(w_{y1} - w_{y2}).$$

Again, referring to the velocity diagram, $w_{y1} - w_{y2} = c_{y2} - c_{y1}$ and

$$(p_3 - p_1)/\rho = U(c_{y2} - c_{y1}) = h_3 - h_1. \tag{5.26}$$

It is noted that, for an isentropic process, $Tds = 0 = dh - (1/\rho)dp$ and therefore, $\Delta h = (1/\rho)\Delta p$.

The pressure rise in a real stage (involving irreversible processes) can be determined if the stage efficiency is known. Defining the stage efficiency η_s as the ratio of the isentropic enthalpy rise to the actual enthalpy rise corresponding to the same *finite* pressure change, (cf. Fig. 2.7), this can be written as

$$\eta_s = (\Delta h_{is})/(\Delta h) = (1/\rho)\Delta p/\Delta h.$$

Thus,

$$(1/\rho)\Delta p = \eta_s \Delta h = \eta_s\, U\Delta c_y. \tag{5.27}$$

If $c_1 = c_3$, then η_s is a very close approximation of the total-to-total efficiency η_{tt}. Although the above expressions are derived for incompressible flow they are, nevertheless, a valid approximation for compressible flow if the stage temperature (and pressure) rise is small.

PRESSURE RATIO OF A MULTISTAGE COMPRESSOR

It is possible to apply the preceding analysis to the determination of multistage compressor pressure ratios. The procedure requires the calculation of pressure and temperature changes for a single stage, the stage exit conditions enabling the density at entry to the following stage to be found. This calculation is repeated for each stage in turn until the required final conditions are satisfied. However, for compressors having identical stages it is more convenient to resort to a simple compressible flow analysis. An illustrative example is given below.

EXAMPLE: A multistage axial compressor is required for compressing air at 293 K, through a pressure ratio of 5 to 1. Each stage is to be 50% reaction and the mean blade speed 275 m/s, flow coefficient 0·5, and stage loading factor 0·3, are taken, for simplicity, as constant for all stages. Determine the flow angles and, the number of stages required if the stage efficiency is 88·8%. Take $C_p = 1·005$ kJ/(kg °C) and $\gamma = 1·4$ for air.

From eqn. (5.14a) the stage load factor can be written as,

$$\psi = \phi(\tan \beta_1 - \tan \beta_2).$$

From eqn. (5.11) the reaction is

$$R = \frac{\phi}{2} (\tan \beta_1 + \tan \beta_2).$$

Solving for $\tan \beta_1$ and $\tan \beta_2$ gives

$$\tan \beta_1 = (R + \psi/2)/\phi \quad \text{and} \quad \tan \beta_2 = (R - \psi/2)/\phi.$$

Calculating β_1 and β_2 and observing for $R = 0·5$ that the velocity diagram is symmetrical,

$$\beta_1 = \alpha_2 = 52·45 \text{ deg} \quad \text{and} \quad \beta_2 = \alpha_1 = 35 \text{ deg}.$$

Writing the stage load factor as $\psi = C_p \Delta T_0 / U^2$, then the stage stagnation temperature rise is,

$$\Delta T_0 = \psi U^2 / C_p = 0·3 \times 275^2 / 1005 = 22·5°C.$$

It is reasonable to take the stage efficiency as equal to the polytropic efficiency since the stage temperature rise of an axial compressor is small. Denoting compressor inlet and outlet conditions by subscripts I and II respectively then, from eqn. (2.33),

$$\frac{T_{0II}}{T_{0I}} = 1 + \frac{N \Delta T_0}{T_{0I}} = \left(\frac{p_{0II}}{p_{0I}} \right)^{(\gamma - 1)/\eta_p \gamma},$$

where N is the required number of stages. Thus

$$N = \frac{T_{01}}{\Delta T_0}\left[\left(\frac{p_{011}}{p_{01}}\right)^{(\gamma-1)/n_p\gamma} - 1\right] = \frac{293}{22\cdot5}[5^{1/3\cdot11} - 1] = 8\cdot86.$$

A suitable number of stages is therefore 9.

The overall efficiency is found from eqn. (2.36).

$$\eta_{tt} = \left[\left(\frac{p_{011}}{p_{01}}\right)^{(\gamma-1)/\gamma} - 1\right]\Big/\left[\left(\frac{p_{011}}{p_{01}}\right)^{(\gamma-1)/n_p\gamma} - 1\right]$$

$$= [5^{1/3\cdot5} - 1]/[5^{1/3\cdot11} - 1] = 86\cdot3\%.$$

ESTIMATION OF COMPRESSOR STAGE EFFICIENCY

In eqn. (5.9) the amount of the actual stage work $(h_{03}-h_{01})$ can be found from the velocity diagram. The losses in total pressure may be estimated from cascade data. This data is incomplete however, as it only takes account of the blade profile loss. Howell[2] has subdivided the total losses into three categories as shown in Fig. 3.11.

  (i) Profile losses on the blade surfaces.
 (ii) Skin friction losses on the annulus walls.
(iii) "Secondary" losses by which he meant all losses not included in (i) and (ii) above.

In performance estimates of axial compressor and fan stages the *overall* drag coefficient for the blades of each row is obtained from

$$C_D = C_{Dp} + C_{Da} + C_{Ds}$$

$$= C_{Dp} + 0\cdot02\, s/H + 0\cdot018\, C_L^2 \tag{5.28}$$

using the empirical values given in Chapter 3.

Although the subject matter of this chapter is primarily concerned with two-dimensional flows, there is an interesting three-dimensional aspect which cannot be ignored. In multistage axial compressors the annulus wall boundary layers rapidly thicken through the first few stages and the axial velocity profile becomes increasingly peaked. This effect is illustrated in Fig. 5.7, from the experimental results of Howell,[2] which shows axial velocity traverses through a four-stage compressor.

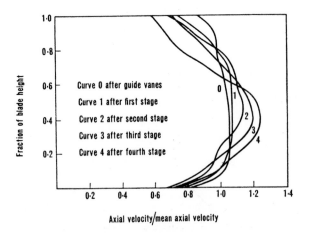

Fig. 5.7. Axial velocity profiles in a compressor (Howell[2]). (Courtesy of the Institution of Mechanical Engineers.)

Over the central region of the blade, the axial velocity is higher than the mean value based on the through flow. The mean blade section (and most of the span) will, therefore, do less work than is estimated from the velocity triangles based on the mean axial velocity. In theory it would be expected that the tip and root sections would provide a compensatory effect because of the low axial velocity in these regions. Due to stalling of these sections (and tip leakage) no such work increase actually occurs, and the net result is that the work done by the whole blade is below the design figure. Howell[2] suggested that the stagnation enthalpy rise across a stage could be expressed as

$$h_{03} - h_{01} = \lambda U(c_{y2} - c_{y1}), \tag{5.29}$$

where λ is a "work-done factor". For multistage compressors Howell recommended for λ a mean value of 0·86. Using a similar argument for axial turbines, the increase in axial velocity at the pitch-line gives an *increase* in the amount of work done, which is then roughly cancelled out by the loss in work at the blade ends. Thus, for turbines, no "work-done factor" is required as a correction in performance calculations.

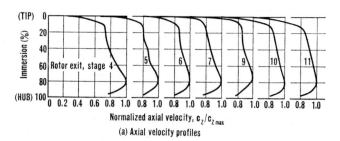

(a) Axial velocity profiles

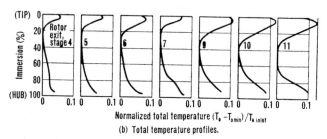

(b) Total temperature profiles.

FIG. 5.8. Traverse measurements obtained from a 12-stage compressor (Smith[15]). (Courtesy of the Elsevier Publishing Co.)

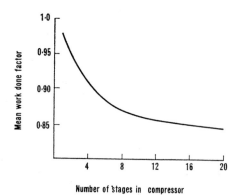

Number of stages in compressor

FIG. 5.9. Mean work done factor in compressors (Howell and Bonham[6]). (Courtesy of the Institution of Mechanical Engineers.)

Other workers have suggested that λ should be high at entry (0·96) where the annulus wall boundary layers are thin, reducing progressively in the later stages of the compressor (0·85). Howell[6] has given mean "work-done factors" for compressors with varying numbers of stages, as in Fig. 5.9. For a four-stage compressor the value of λ would be 0·9 which would be applied to all four stages.

More recently, Smith[15] has commented upon the rather pronounced deterioration of compressor performance implied by the example given in Fig. 5.7 and suggested that things are not so bad as suggested. As an example of modern practice he gave the axial velocity distributions through a twelve-stage axial compressor, Fig. 5.8(a). This does illustrate that rapid changes in velocity distribution still occur in the first few stages, but that the profile settles down to a fairly constant shape thereafter. This phenomenon has been referred to as *ultimate steady flow*.

Smith also provided curves of the spanwise variation in total temperature, Fig. 5.8(b), which shows the way losses increase from mid-passage towards the annulus walls. An inspection of this figure shows also that the excess total temperature near the end walls increases in magnitude and extent as the flow passes through the compressor. Work on methods of predicting annulus wall boundary layers in turbomachines and their effects on performance are being actively pursued in many countries. Although rather beyond the scope of this textbook, it may be worth mentioning two fairly recent papers for students wishing to advance their studies further. Mellor and Balsa[16] offer a mathematical flow model based on the pitchwise-averaged turbulent equations of motion for predicting axial flow compressor performance whilst Daneshyar et al.[17] review and compare different existing methods for predicting the growth of annulus wall boundary layers in turbomachines.

EXAMPLE: The last stage of an axial flow compressor has a reaction of 50% at the design operating point. The cascade characteristics, which correspond to each row at the mean radius of the stage, are shown in Fig. 3.10. These apply to a cascade of circular arc camber line blades at a space–chord ratio 0·9, a blade inlet angle of 44·5 deg and a blade outlet angle of −0·5 deg. The blade height–chord ratio is 2·0 and the work done factor can be taken as 0·86 when the mean radius relative incidence $(i-i^*)/\varepsilon^*$ is 0·4 (the operating point).

For this operating condition, determine

    (i) the nominal incidence i^* and nominal deflection ε^*;
    (ii) the inlet and outlet flow angles for the rotor;
    (iii) the flow coefficient and stage loading factor;
    (iv) the rotor lift coefficient;
    (v) the overall drag coefficient of each row;
    (vi) the stage efficiency.

The density at entrance to the stage is 3·5 kg/m³ and the mean radius blade speed is 242 m/s. Assuming the density across the stage is constant and ignoring compressibility effects, estimate the stage pressure rise.

In the solution given below the *relative flow* onto the rotor is considered. The notation used for flow angles is the same as for Fig. 5.2. For blade angles, β' is therefore used instead of α' for the sake of consistency.

(i) The nominal deviation is found using eqns (3.39) and (3.40). With the camber $\theta = \beta_1' - \beta_2' = 44\cdot5° - (-0\cdot5°) = 45°$ and the space/chord ratio, $s/l = 0\cdot9$, then

$$\delta^* = [0\cdot23 + \beta_2^*/500]\theta(s/l)^{\frac{1}{2}}$$

But
$$\beta_2^* = \delta^* + \beta_2' = \delta^* - 0\cdot5$$

$$\therefore\ \delta^* = [0\cdot23 + (\delta^* + \beta_2')/500] \times 45 \times (0\cdot9)^{\frac{1}{2}}$$

$$= [0\cdot229 + \delta^*/500] \times 42\cdot69 = 9\cdot776 + 0\cdot0854\,\delta^*$$

$$\therefore\delta^* = 10\cdot69°$$

$$\therefore\beta_2^* = \delta^* + \beta_2' = 10\cdot69 - 0\cdot5$$

$$\simeq 10\cdot2°$$

The nominal deflection $\varepsilon^* = 0\cdot8\,\varepsilon_{max}$ and, from Fig. 3.10, $\varepsilon_{max} = 37\cdot5°$. Thus, $\varepsilon^* = 30°$ and the nominal incidence is

$$i^* = \beta_2^* + \varepsilon^* - \beta_1'$$
$$= 10\cdot2 + 30 - 44\cdot5 = -4\cdot3°.$$

(ii) At the operating point $i = 0\cdot4\,\varepsilon^* + i^* = 7\cdot7°$. Thus, the actual inlet flow angle is

$$\beta_1 = \beta_1' + i = 52\cdot2°.$$

From Fig. 3.10 at $i = 7 \cdot 7°$, the deflection $\varepsilon = 37 \cdot 5°$ and the flow outlet angle is

$$\beta_2 = \beta_1 - \varepsilon = 14 \cdot 7°.$$

(iii) From Fig. 5.2, $U = c_{x1} (\tan \alpha_1 + \tan \beta_1) = c_{x2}(\tan \alpha_2 + \tan \beta_2)$. For $c_x = $ constant across the stage and $R = 0 \cdot 5$

$$\beta_1 = \alpha_2 = 52 \cdot 2° \quad \text{and} \quad \beta_2 = \alpha_1 = 14 \cdot 7°$$

and the flow coefficient is

$$\phi = \frac{c_x}{U} = \frac{1}{\tan \alpha_1 + \tan \beta_1} = 0 \cdot 644.$$

The stage loading factor, $\psi = \Delta h_0 / U^2 = \lambda \phi (\tan \alpha_2 - \tan \alpha_1)$ using eqn. (5.29). Thus, with $\lambda = 0 \cdot 86$,

$$\psi = 0 \cdot 568.$$

(iv) The lift coefficient can be obtained using eqn. (3.18)

$$C_L = 2(s/l) \cos \beta_m (\tan \beta_1 - \tan \beta_2)$$

ignoring the small effect of the drag coefficient. Now $\tan \beta_m = (\tan \beta_1 + \tan \beta_2)/2$. Hence $\beta_m = 37 \cdot 8°$ and so

$$C_L = 2 \times 0 \cdot 9 \times 0 \cdot 7902 \times 1 \cdot 027 = 1 \cdot 46.$$

(v) Combining eqns. (3.7) and (3.17) the drag coefficient is

$$C_D = \frac{s}{l} \left(\frac{\Delta p_0}{\frac{1}{2} \rho w_1^2} \right) \frac{\cos^3 \beta_m}{\cos^2 \beta_1}.$$

Again using Fig. 3.10 at $i = 7 \cdot 7°$, the profile total pressure loss coefficient $\Delta p_0 / (\frac{1}{2} \rho w_1^2) = 0 \cdot 032$, hence the profile drag coefficient for the blades of either row is

$$C_{Dp} = 0 \cdot 9 \times 0 \cdot 032 \times 0 \cdot 7902^3 / 0 \cdot 6129^2 = 0 \cdot 038.$$

Taking account of the annulus wall drag coefficient C_{Da} and the secondary loss drag coefficient C_{Ds}

$$C_{Da} = 0 \cdot 02(s/l)(l/H) = 0 \cdot 02 \times 0 \cdot 9 \times 0 \cdot 5 = 0 \cdot 009$$
$$C_{Ds} = 0 \cdot 018 C_L^2 = 0 \cdot 018 \times 1 \cdot 46^2 = 0 \cdot 038.$$

Thus the *overall* drag coefficient, $C_D = C_{Dp} + C_{Da} + C_{Ds} = 0.084$ and this applies to each row of blades. If the reaction had been other than 0.5 the drag coefficients for each blade row would have been computed separately.

(vi) The total-to-total stage efficiency, using eqn. (5.9) can be written as

$$\eta_{tt} = 1 - \frac{\Sigma \Delta p_0 / \rho}{\psi U^2} = 1 - \frac{\Sigma \Delta p_0 / (\frac{1}{2} \rho c_x{}^2)}{2\psi / \phi^2} = 1 - \frac{(\zeta_R + \zeta_S)\phi^2}{2\psi}$$

where ζ_R and ζ_S are the overall total pressure loss coefficients for the rotor and stator rows respectively. From eqn. (3.17)

$$\zeta_s = (l/s)C_D \sec^3 \alpha_m.$$

Thus, with $\zeta_S = \zeta_R$

$$\eta_{tt} = 1 - \frac{\phi^2 C_D(l/s)}{\psi \cos^3 \alpha_m}$$

$$= 1 - \frac{0.644^2 \times 0.084}{0.568 \times 0.7903^3 \times 0.9} = 0.862.$$

From eqn. (5.27), the pressure rise across the stage is

$$\Delta p = \eta_{tt} \, \psi \rho U^2 = 0.862 \times 0.568 \times 3.5 \times 242^2$$

$$= 100 \text{ kPa}.$$

STABILITY OF COMPRESSORS

A salient feature of a compressor performance map, such as Fig. 1.9, is the limit to stable operation known as the *surge line*. This limit is reached by reducing the mass flow (with a throttle valve) whilst the rotational speed is maintained constant.

When a compressor goes into surge the effects are usually quite dramatic. Generally, an increase in noise level is experienced, indicative of a pulsation of the air flow and of mechanical vibration. Commonly, there are a small number of predominant frequencies superimposed on a high background noise. The lowest frequencies are usually associated with a *Helmholtz-type of resonance* of the flow through the machine, with the inlet and/or outlet volumes. The higher frequencies are known to be due to *rotating stall* and are of the same order as the rotational speed of the impeller.

Rotating stall is a phenomenon of axial-compressor flow which has been the subject of many detailed experimental and theoretical investigations and the matter is still not fully resolved. A survey of the subject is given by Emmons *et al.*[7] and one of several theoretical treatments in ref. 8. Briefly, when a blade row (usually the rotor) of a compressor reaches the "stall point", the blades instead of all stalling together as might be expected, stall in separate patches and these stall patches, moreover, travel around the compressor annulus (i.e. they rotate).

That stall patches *must* propagate from blade to blade has a simple physical explanation. Consider a portion of a blade row, as illustrated in Fig. 5.10 to be affected by a stall patch. This patch must cause a partial obstruction to the flow which is deflected on both sides of it.

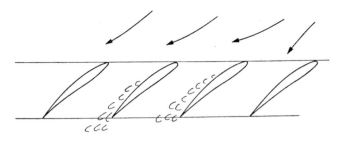

FIG. 5.10. Model illustrating mechanism of stall cell propagation: partial blockage due to stall patch deflects flow, increasing incidence to the left and decreasing incidence to the right.

Thus, the incidence of the flow on to the blades on the right of the stall cell is reduced but, the incidence to the left is increased. As these blades are already close to stalling, the net effect is for the stall patch to move to the left; the motion is then self-sustaining.

There is a strong practical reason for the wide interest in rotating stall. Stall patches travelling around blade rows load and unload each blade at some frequency related to the speed and number of the patches. This frequency may be close to a natural frequency of blade vibration and there is clearly a need for accurate prediction of the conditions producing such a vibration. Several cases of blade failure due to resonance induced by rotating stall have been reported, usually with serious consequences to the whole compressor.

It is possible to distinguish between surge and propagating stall by the unsteadiness, or otherwise, of the total mass flow. The characteristic of stall propagation is that the flow passing through the annulus, summed over the whole area, is steady with time; the stall cells merely redistribute the flow over the annulus. Surge, on the other hand, involves an axial oscillation of the total mass flow, a condition highly detrimental to efficient compressor operation.

The conditions determining the point of surge of a compressor have not yet been completely determined satisfactorily. One physical explanation of this breakdown of the flow is given by Horlock.[5]

Figure 5.11 shows a constant rotor speed compressor characteristic (C) of pressure ratio plotted against flow coefficient. A second set of curves (T_1, T_2, etc.) are superimposed on this figure showing the

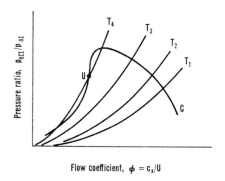

FIG. 5.11. Stability of operation of a compressor (adapted from Horlock[5]).

pressure loss characteristics of the throttle for various fixed throttle positions. The intersection of curves T with compressor curve C denotes the various operating points of the combination. A state of *flow stability* exists if the throttle curves at the point of intersection have a greater (positive) slope than the compressor curve. That this is so may be illustrated as follows. Consider the operating point at the intersection of T_2 with C. If a small reduction of flow should momentarily occur, the compressor will produce a greater pressure rise and the throttle resistance will fall. The flow rate must, perforce, increase so that the original operating point is restored. A similar argument holds if the flow is temporarily augmented, so that the flow is completely stable at this operating condition.

If now the operating point is at point U unstable operation is possible. A small reduction in flow will cause a greater reduction in compressor pressure ratio than the corresponding pressure ratio across the throttle. As a consequence of the increased resistance of the throttle, the flow will decrease even further and the operating point U is clearly unstable. By inference, neutral stability exists when the slopes of the throttle pressure loss curves equal the compressor pressure rise curve.

Tests on low pressure ratio compressors appear to substantiate this explanation of instability. However, for high rotational speed multi-stage compressors the above argument does not seem sufficient to describe surging. With high speeds no stable operation appears possible on constant speed curves of positive slope and surge appears to occur when this slope is zero or even a little negative. A more complete understanding of surge in multistage compressors is only possible from a detailed study of the individual stages performance and their inter-action upon one another.

AXIAL-FLOW DUCTED FANS

In essence, an axial-flow fan is simply a single-stage compressor of low pressure (and temperature) rise, so that much of the foregoing theory of this chapter is valid for this class of machine. However, because of the high space–chord ratio used in many axial fans, a simplified theoretical approach based on *isolated aerofoil theory* is often used. This method can be of use in the design of ventilating fans (usually of high space–chord) in which aerodynamic interference between adjacent blades can be assumed negligible. Attempts have been made to extend the scope of isolated aerofoil theory to less widely spaced blades by the introduction of an *interference factor*; for instance, the ratio k of the lift force of a single blade in a cascade to the lift force of a single isolated blade. As a guide to the degree of this interference, the exact solution obtained by Weinig[9] and used by Wislicenus[10] for a row of thin flat plates is of value and is shown in Fig. 5.12. This

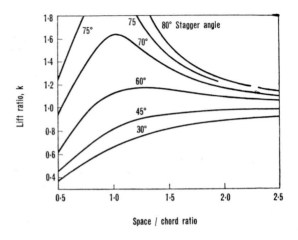

Space / chord ratio

FIG. 5.12. Weinig's results for lift ratio of a cascade of thin flat plates, showing dependence on stagger angle and space/chord ratio (adapted from Wislicenus[10]).

illustrates the dependence of k on space–chord ratio for several stagger angles. The rather pronounced effect of stagger for moderate space–chord ratios should be noted as well as the asymptotic convergence of k towards unity for higher space–chord ratios.

Two simple types of axial-flow fan are shown in Fig. 5.13 in which the inlet and outlet flows are entirely axial. In the first type (a), a set of guide vanes provide a contra-swirl and the flow is restored to the axial direction by the rotor. In the second type (b), the rotor imparts swirl in the direction of blade motion and the flow is restored to the axial direction by the action of outlet *straighteners* (or outlet guide vanes). The theory and design of both the above types of fan have been investigated by Van Niekerk[11] who was able to formulate expressions for calculating the optimum sizes and fan speeds using blade element theory.

BLADE ELEMENT THEORY

A blade element at a given radius can be defined as an aerofoil of vanishingly small span. In fan-design theory it is commonly assumed that each such element operates as a two-dimensional aerofoil, behaving completely independently of conditions at any other radius. Now

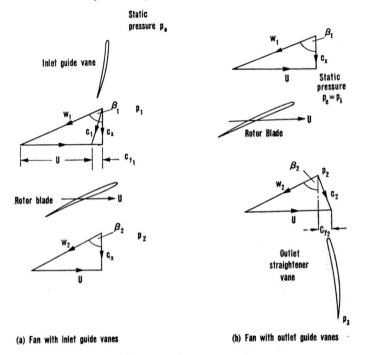

(a) Fan with inlet guide vanes (b) Fan with outlet guide vanes

FIG. 5.13. Two simple types of axial-flow fan and their associated velocity
diagrams (after Van Niekerk[11]).

the forces impressed upon the fluid by unit span of a single stationary
blade have been considered in some detail already, in Chapter 3.
Considering an *element of* a *rotor* blade dr, at radius r, the elementary
axial and tangential forces, dX and dY respectively, exerted on the
fluid are, referring to Fig. 3.5,

$$dX = (L \sin \beta_m - D \cos \beta_m)dr, \tag{5.30}$$

$$dY = (L \cos \beta_m + D \sin \beta_m)dr, \tag{5.31}$$

where $\tan \beta_m = \frac{1}{2}\{\tan \beta_1 + \tan \beta_2\}$ and L, D are the lift and drag on
unit span of a blade.

Writing $\tan \gamma = D/L = C_D/C_L$ then,

$$dX = L(\sin \beta_m - \tan \gamma \cos \beta_m)dr.$$

Introducing the lift coefficient $C_L = L/(\frac{1}{2}\rho w_m^2 l)$ for the rotor blade (cf. eqn. (3.16a)) into the above expression and rearranging,

$$dX = \frac{\rho c_x^2 l C_L dr}{2 \cos^2 \beta_m} \cdot \frac{\sin(\beta_m - \gamma)}{\cos \gamma},\tag{5.32}$$

where $c_x = w_m \cos \beta_m$.

The torque exerted by *one* blade element at radius r is rdY. If there are Z blades the elementary torque is

$$d\tau = rZdY$$

$$= rZL(\cos \beta_m + \tan \gamma \sin \beta_m)dr,$$

after using eqn. (5.31). Substituting for L and rearranging,

$$d\tau = \frac{\rho c_x^2 l Z C_L r dr}{2 \cos^2 \beta_m} \cdot \frac{\cos(\beta_m - \gamma)}{\cos \gamma}.\tag{5.33}$$

Now the work done by the rotor in unit time equals the product of the stagnation enthalpy rise and the mass flow rate; for the elementary ring of area $2\pi rdr$,

$$\Omega d\tau = (C_p \Delta T_0)d\dot{m},\tag{5.34}$$

where Ω is the rotor angular velocity and the element of mass flow, $d\dot{m} = \rho c_x 2\pi rdr$.

Substituting eqn. (5.33) into eqn. (5.34), then

$$C_p \Delta T_0 = C_p \Delta T = C_L \frac{Uc_x}{2} \frac{l \cos(\beta_m - \gamma)}{s \cos^2 \beta_m \cos \gamma}.\tag{5.35}$$

where $s = 2\pi r/Z$. Now the static temperature rise equals the stagnation temperature rise when the velocity is unchanged across the fan; this, in fact, is the case for both types of fan shown in Fig. 5.13.

The increase in static pressure of the *whole* of the fluid crossing the rotor row may be found by equating the total axial force on all the blade elements at radius r with the product of static pressure rise and elementary area $2\pi rdr$, or

$$ZdX = (p_2 - p_1)2\pi rdr.$$

Using eqn. (5.32) and rearranging,

$$p_2 - p_1 = C_L \frac{\rho c_x^2}{2} \frac{l}{s} \frac{\sin(\beta_m - \gamma)}{\cos^2\beta_m \cos\gamma} \qquad (5.36)$$

Note that, so far, all the above expressions are applicable to boih types of fan shown in Fig. 5.13.

BLADE ELEMENT EFFICIENCY

Consider the fan type shown in Fig. 5.13a fitted with guide vanes at inlet. The pressure rise across this fan is equal to the rotor pressure rise (p_2-p_1) *minus* the drop in pressure across the guide vanes (p_e-p_1). The ideal pressure rise across the fan is given by the product of density and $C_p\Delta T_0$. Fan designers define a blade element efficiency

$$\eta_b = \{(p_2 - p_1) - (p_e - p_1)\}/(\rho C_p\Delta T_0). \qquad (5.37)$$

The drop in static pressure across the guide vanes, assuming *frictionless* flow for simplicity, is

$$p_e - p_1 = \tfrac{1}{2}\rho(c_1^2 - c_x^2) = \tfrac{1}{2}\rho c_{y1}^2. \qquad (5.38)$$

Now since the change in swirl velocity across the rotor is equal and opposite to the swirl produced by the guide vanes, the work done per unit mass flow, $C_p\Delta T_0$ is equal to Uc_{y1}. Thus the second term in eqn. (5.37) is

$$(p_e - p_1)/(\rho C_p\Delta T_0) = c_{y1}/(2U). \qquad (5.39)$$

Combining eqns. (5.35), (5.36) and (5.39) in eqn. (5.37), then

$$\eta_b = (c_x/U)\tan(\beta_m - \gamma) - c_{y1}/(2U). \qquad (5.40a)$$

The foregoing exercise can be repeated for the second type of fan having outlet straightening vanes and, assuming frictionless flow through the "straighteners", the rotor blade element efficiency becomes,

$$\eta_b = (c_x/U)\tan(\beta_m-\gamma) + c_{y2}/(2U). \qquad (5.40b)$$

Some justification for ignoring the losses occurring in the guide vanes is found by observing that the ratio of guide vane pressure change to rotor pressure rise is normally small in ventilating fans. For example, in the first type of fan

$$(p_e - p_1)/(p_2 - p_1) \doteq (\tfrac{1}{2}\rho c_{y1}^2)/(\rho U c_{y1}) = c_{y1}/2(U),$$

the tangential velocity c_{y1} being rather small compared with the blade speed U.

LIFT COEFFICIENT OF A FAN AEROFOIL

For a specified blade element geometry, blade speed and lift/drag ratio the temperature and pressure rises can be determined if the lift coefficient is known. An estimate of lift coefficient is most easily obtained from two-dimensional aerofoil potential flow theory. Glauert[12] shows for isolated aerofoils of small camber and thickness, that

$$C_L = 2\pi \sin \chi, \qquad (5.41)$$

where χ is the angle between the flow direction and *line of zero lift* of the aerofoil. For an isolated, cambered aerofoil Wislicenus[10]

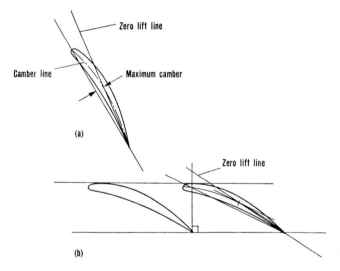

FIG. 5.14. Method suggested by Wislicenus[10] for obtaining the zero lift line of cambered aerofoils.

suggested that the zero lift line may be found by joining the trailing edge point with the point of maximum camber as depicted in Fig. 5.14a. For fan blades experiencing some interference effects from adjacent blades, the modified lift coefficient of a blade may be estimated by assuming that Weinig's results for flat plates (Fig. 5.12) are valid for the slightly cambered, finite thickness blades, and

$$C_L = 2\pi k \sin \chi. \tag{5.41a}$$

When the vanes overlap (as they may do at sections close to the hub), Wislicenus suggested that the zero lift line may be obtained by the line connecting the trailing edge point with the maximum camber of that portion of blade which is not overlapped (Fig. 5.14b).

The extension of both blade element theory and cascade data to the design of complete fans is discussed in considerable detail by Wallis.[13]

REFERENCES

1. STONEY, G., Scientific activities of the late Hon. Sir Charles Parsons, F.R.S. *Engineering, London*, **144** (1937).
2. HOWELL, A. R., Fluid dynamics of axial compressors. *Proc. Instn. Mech. Engrs. London*, **153** (1945).
3. COX, H. ROXBEE, British aircraft gas turbines. *J. Aero. Sci.* **13** (1946).
4. CONSTANT, H., The early history of the axial type of gas turbine engine. *Proc. Instn. Mech. Engrs. London*, **153** (1945).
5. HORLOCK, J. H., *Axial Flow Compressors*. Butterworths, London (1958).
6. HOWELL, A. R. and BONHAM, R. P., Overall and stage characteristics of axial flow compressors. *Proc. Instn. Mech. Engrs. London*, **163** (1950).
7. EMMONS, H. W., KRONAUER, R. E. and ROCKET, J. A., A survey of stall propagation—experiment and theory. *Trans. Am. Soc. Mech. Engrs.* Series D, **81** (1959).
8. DIXON, S. L., Some three-dimensional effects of rotating stall. A.R.C. Current Paper No. 609 (1962).
9. WEINIG, F., *Die Stroemung um die Schaufeln von Turbomaschinen*, Joh. Ambr. Barth, Leipzig (1935).
10. WISLICENUS, G. F., *Fluid Mechanics of Turbomachinery*. McGraw-Hill, New York (1947).
11. VAN NIEKERK, C. G., Ducted fan design theory. *J. Appl. Mech.* **25** (1958).
12. GLAUERT, H., *The Elements of Aerofoil and Airscrew Theory*. Cambridge University Press (1959).
13. WALLIS, R. A., *Axial Flow Fans, Design and Practice*. Newnes, London (1961).
14. GOSTELOW, J. P., HORLOCK, J. H. and MARSH, H., Recent developments in the aerodynamic design of axial flow compressors. Symposium at Warwick University. *Proc. Instn. Mech. Engrs. London*, **183**, Pt. 3N (1969).

15. SMITH, L. H., Jr., Casing boundary layers in multistage compressors. *Proc. Symposium on Flow Research on Blading*. Ed. L. S. DZUNG, Elsevier (1970).
16. MELLOR, G. L. and BALSA, T. F., The prediction of axial compressor performance with emphasis on the effect of annulus wall boundary layers. *Agardograph No. 164*. Advisory Group for Aerospace Research and Development (1972).
17. DANESHYAR, M., HORLOCK, J. H. and MARSH, H., Prediction of annulus wall boundary layers in axial flow turbomachines. *Agardograph No. 164*. Advisory Group for Aerospace Research and Development (1972).

PROBLEMS

(*Note*: In questions 1 to 4 take $R = 287 \text{J}/(\text{kg }^{\circ}\text{C})$ and $\gamma = 1\cdot4$.)

1. An axial flow compressor is required to deliver 50 kg/s of air at a stagnation pressure of 500 kPa. At inlet to the first stage the stagnation pressure is 100 kPa and the stagnation temperature is 23°C. The hub and tip diameters at this location are 0·436 m and 0·728 m. At the mean radius, which is constant through all stages of the compressor, the reaction is 0·50 and the absolute air angle at stator exit is 28·8 deg for all stages. The speed of the rotor is 8000 rev/min. Determine the number of similar stages needed assuming that the polytropic efficiency is 0·89 and that the axial velocity at the mean radius is constant through the stages and equal to 1·05 times the average axial velocity.

2. Derive an expression for the degree of reaction of an axial compressor stage in terms of the flow angles relative to the rotor and the flow coefficient.

Data obtained from early cascade tests suggested that the limit of efficient working of an axial-flow compressor stage occurred when

  (i) a *relative* Mach number of 0·7 on the rotor is reached;
  (ii) the flow coefficient is 0·5;
  (iii) the relative flow angle at rotor outlet is 30 deg measured from the axial direction;
  (iv) the stage reaction is 50%.

Find the limiting stagnation temperature rise which would be obtained in the first stage of an axial compressor working under the above conditions and compressing air at an inlet *stagnation* temperature of 289 K. Assume the axial velocity is constant across the stage.

3. Each stage of an axial flow compressor is of 0·5 reaction, has the same mean blade speed and the same flow outlet angle of 30 deg relative to the blades. The mean flow coefficient is constant for all stages at 0·5. At entry to the first stage the stagnation temperature is 278 K, the stagnation pressure 101·3 kPa, the static pressure is 87·3 kPa and the flow area 0·372 m². Using compressible flow analysis determine the axial velocity and the mass flow rate.

Determine also the shaft power needed to drive the compressor when there are 6 stages and the mechanical efficiency is 0·99.

4. A sixteen-stage axial flow compressor is to have a pressure ratio of 6·3. Tests have shown that a stage total-to-total efficiency of 0·9 can be obtained for each of the first six stages and 0·89 for each of the remaining ten stages. Assuming constant work done in each stage and similar stages find the compressor overall total-to-total efficiency. For a mass flow rate of 40 kg/s determine the power required by the compressor. Assume an inlet total temperature of 288 K.

5. At a particular operating condition an axial flow compressor has a reaction of 0·6, a flow coefficient of 0·5 and a stage loading, defined as $\Delta h_0/U^2$ of 0·35. If the flow exit angles for each blade row may be assumed to remain unchanged when the mass flow is throttled, determine the reaction of the stage and the stage loading when the air flow is reduced by 10% at constant blade speed. Sketch the velocity triangles for the two conditions.

Comment upon the likely behaviour of the flow when further reductions in air mass flow are made.

6. The proposed design of a compressor rotor blade row is for 59 blades with a circular arc camber line. At the mean radius of 0·254 m the blades are specified with a camber of 30 deg, a stagger of 40 deg and a chord length of 30 mm. Determine, using Howell's correlation method, the nominal outlet angle, the nominal deviation and the nominal inlet angle. The tangent difference approximation, proposed by Howell for nominal conditions ($0 \leqslant a_2 {}^* \leqslant 40°$), can be used:

$$\tan a_1{}^* - \tan a_2{}^* = 1·55/(1 + 1·5\, s/l).$$

Determine the nominal lift coefficient given that the blade drag coefficient $C_D = 0·017$.

Using the data for relative deflection given in Fig. 3.17, determine the flow outlet angle and lift coefficient when the incidence $i = 1·8$ deg. Assume that the drag coefficient is unchanged from the previous value.

7. The preliminary design of an axial flow compressor is to be based upon a simplified consideration of the mean diameter conditions. Suppose that the stage characteristics of a repeating stage of such a design are as follows:

Stagnation temperature rise	25°C
Reaction ratio	0·6
Flow coefficient	0·5
Blade speed	275 m/s

The gas compressed is air with a specific heat at constant pressure of 1·005 kJ/(kg°C). Assuming constant axial velocity across the stage and equal absolute velocities at inlet and outlet, determine the relative flow angles for the rotor.

Physical limitations for this compressor dictate that the space/chord ratio is unity at the mean diameter. Using Howell's correlation method, determine a suitable camber at the mid-height of the rotor blades given that the incidence angle is zero. Use the tangent difference approximation

$$\tan \beta_1{}^* - \tan \beta_2{}^* = 1·55/(1 + 1·5\, s/l)$$

for nominal conditions and the data of Fig. 3.17 for finding the design deflection. (*Hint.* Use several trial values of θ to complete the solution.)

CHAPTER 6

Three-dimensional Flows in Axial Turbomachines

It cost much labour and many days before all these things were brought to perfection.
(DEFOE, *Robinson Crusoe.*)

IN CHAPTERS 4 and 5 the fluid motion through the blade rows of axial turbomachines was assumed to be two-dimensional in the sense that radial (i.e. spanwise) velocities did not exist. This is a not unreasonable assumption for axial turbomachines of high hub–tip ratio. However, with hub–tip ratios less than about 4/5, radial velocities through a blade row may become appreciable, the consequent redistribution of mass flow (with respect to radius) seriously affecting the outlet velocity profile (and flow angle distribution). It is the temporary imbalance between the strong centrifugal forces exerted on the fluid and radial pressures restoring equilibrium which is responsible for these radial flows. Thus, to an observer travelling with a fluid particle, radial motion will continue until sufficient fluid is transported (radially) to change the pressure distribution to that necessary for equilibrium. The flow in an annular passage in which there is no radial component of velocity, whose streamlines lie in circular, cylindrical surfaces and which is axisymmetric, is commonly known as *radial equilibrium* flow.

An analysis called *the radial equilibrium method*, widely used for three-dimensional design calculations in axial compressors and turbines, is based upon the assumption that any radial flow which may occur, is completed *within* a blade row, the flow *outside* the row then being in radial equilibrium. Figure 6.1 illustrates the nature of this assumption. The other assumption that the flow is axisymmetric implies that the effect of the discrete blades is not transmitted to the flow.

152

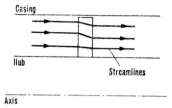

FIG. 6.1. Radial equilibrium flow through a rotor blade row.

THEORY OF RADIAL EQUILIBRIUM

Consider a small element of fluid of mass dm, shown in Fig. 6.2, of unit depth and subtending an angle $d\theta$ at the axis, rotating about the

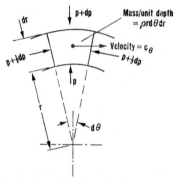

FIG. 6.2. A fluid element in radial equilibrium ($c_r = 0$).

axis with tangential velocity, c_θ at radius r. The element is in radial equilibrium so that the pressure forces balance the centrifugal forces;

$$(p + dp)(r + dr)d\theta - prd\theta - (p + \tfrac{1}{2}dp)drd\theta = dmc_\theta^2/r.$$

Writing $dm = \rho r d\theta dr$ and ignoring terms of the second order of smallness the above equation reduces to,

$$\frac{1}{\rho}\frac{dp}{dr} = \frac{c_\theta^2}{r}. \tag{6.1}$$

If the swirl velocity c_θ and density are known functions of radius, the radial pressure variation along the blade length can be determined,

$$p_{\text{tip}} - p_{\text{root}} = \int_{\text{root}}^{\text{tip}} \rho c_\theta^2 \frac{dr}{r}. \tag{6.2a}$$

For an incompressible fluid

$$p_{\text{tip}} - p_{\text{root}} = \rho \int_{\text{root}}^{\text{tip}} c_\theta^2 \frac{dr}{r}. \tag{6.2b}$$

The stagnation enthalpy is written (with $c_r = 0$)

$$h_0 = h + \tfrac{1}{2}(c_x^2 + c_\theta^2) \tag{6.3}$$

therefore,

$$\frac{dh_0}{dr} = \frac{dh}{dr} + c_x \frac{dc_x}{dr} + c_\theta \frac{dc_\theta}{dr}. \tag{6.4}$$

The thermodynamic relation $T\,ds = dh-(1/\rho)dp$ can be similarly written

$$T \frac{ds}{dr} = \frac{dh}{dr} - \frac{1}{\rho} \frac{dp}{dr}. \tag{6.5}$$

Combining eqns. (6.1), (6.4) and (6.5), eliminating dp/dr and dh/dr, the *radial equilibrium equation* may be obtained,

$$\frac{dh_0}{dr} - T \frac{ds}{dr} = c_x \frac{dc_x}{dr} + \frac{c_\theta}{r} \frac{d}{dr}(rc_\theta). \tag{6.6}$$

If the stagnation enthalpy h_0 and entropy s remain the same at all radii, $dh_0/dr = ds/dr = 0$, eqn. (6.6) becomes,

$$c_x \frac{dc_x}{dr} + \frac{c_\theta}{r} \frac{d}{dr}(rc_\theta) = 0. \tag{6.6a}$$

Equation (6.6a) will hold for the flow between the rows of an adiabatic, reversible (ideal) turbomachine in which rotor rows either deliver or receive equal work at all radii. Now if the flow is incompressible, instead of eqn. (6.3) use $p_0 = p + \tfrac{1}{2}\rho(c_x^2 + c_\theta^2)$ to obtain

$$\frac{1}{\rho}\frac{dp_0}{dr} = \frac{1}{\rho}\frac{dp}{dr} + c_x\frac{dc_x}{dr} + c_\theta\frac{dc_\theta}{dr}. \tag{6.7}$$

Combining eqns. (6.1) and (6.7), then

$$\frac{1}{\rho}\frac{dp_0}{dr} = c_x\frac{dc_x}{dr} + \frac{c_\theta}{r}\frac{d}{dr}(rc_\theta). \tag{6.8}$$

Equation (6.8) clearly reduces to eqn. (6.6a) in a turbomachine in which equal work is delivered at all radii and the total pressure losses across a row are uniform with radius.

Equation (6.6a) may be applied to two sorts of problem as follows: (i) the design (or indirect) problem—in which the tangential velocity distribution is specified and the axial velocity variation is found, or (ii) the direct problem—in which the swirl angle distribution is specified, the axial and tangential velocities being determined.

THE INDIRECT PROBLEM

1. *Free-vortex flow*

This is a flow where the product of radius and tangential velocity remains constant (i.e. $rc_\theta = K$, a constant). The term "vortex-free" might be more appropriate as the vorticity (to be precise we mean *axial* vorticity component) is then zero.

Consider an element of an ideal inviscid fluid rotating about some fixed axis, as indicated in Fig. 6.3. The *circulation* Γ, is defined as the

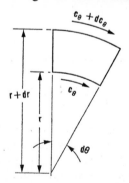

FIG. 6.3. Circulation about an element of fluid.

line integral of velocity around a curve enclosing an area A, or $\Gamma = \oint c \, ds$. The *vorticity* at a point is defined as, the limiting value of circulation $\delta \Gamma$ divided by area δA, as δA becomes vanishingly small. Thus vorticity, $\omega = d\Gamma/dA$.

For the element shown in Fig. 6.3, $c_r = 0$ and

$$d\Gamma = (c_\theta + dc_\theta)(r + dr)d\theta - c_\theta r d\theta$$

$$= \left(\frac{dc_\theta}{dr} + \frac{c_\theta}{r}\right) r d\theta dr$$

ignoring the product of small terms. Thus, $\omega = d\Gamma/dA = (1/r)d(rc_\theta)/dr$. If the vorticity is zero, $d(rc_\theta)/dr$ is also zero and, therefore, rc_θ is constant with radius.

Putting $rc_\theta =$ constant in eqn. (6.6a), then $dc_x/dr = 0$ and so $c_x = $ a constant. This information can be applied to the incompressible flow through a free-vortex compressor or turbine stage, enabling the radial variation in flow angles, reaction and work to be found.

Compressor stage. Consider the case of a compressor stage in which $rc_{\theta 1} = K_1$ before the rotor and $rc_{\theta 2} = K_2$ after the rotor, where K_1, K_2 are constants. The work done by the rotor on unit mass of fluid is

$$\Delta W = U(c_{\theta 2} - c_{\theta 1}) = \Omega r(K_2/r - K_1/r)$$

$$= \text{constant.}$$

Thus, the work done is equal at all radii.

The relative flow angles (see Fig. 5.2) entering and leaving the rotor are

$$\tan \beta_1 = \frac{U}{c_x} - \tan \alpha_1 = \frac{\Omega r - K_1/r}{c_x},$$

$$\tan \beta_2 = \frac{U}{c_x} - \tan \alpha_2 = \frac{\Omega r - K_2/r}{c_x}.$$

in which $c_{x1} = c_{x2} = c_x$ for incompressible flow.

In Chapter 5, reaction in an axial compressor is defined by

$$R = \frac{\text{static enthalpy rise in the rotor}}{\text{static enthalpy rise in the stage}}.$$

For a normal stage ($\alpha_1 = \alpha_3$) with c_x constant across the stage, the reaction was shown to be

$$R = \frac{c_x}{2U}(\tan \beta_1 + \tan \beta_2). \tag{5.11}$$

Substituting values of $\tan \beta_1$ and $\tan \beta_2$ into eqn. (5.11), the reaction becomes

$$R = 1 - \frac{k}{r^2}, \tag{6.9}$$

where

$$k = (K_1 + K_2)/(2\Omega).$$

It will be clear that as k is positive, the reaction increases from root to tip. Likewise, from eqn. (6.1) we observe that as c_θ^2/r is always positive (excepting $c_\theta = 0$), so static pressure increases from root to tip. For the free-vortex flow $rc_\theta = K$, the static pressure variation is obviously $p/\rho = \text{constant} - K/(2r^2)$ upon integrating eqn. (6.1).

EXAMPLE: An axial flow compressor stage is designed to give free-vortex tangential velocity distributions for all radii before and after the rotor blade row. The tip diameter is constant and 1·0 m; the hub diameter is 0·9 m and constant for the stage. At the rotor tip the flow angles are as follows

Absolute inlet angle, α_1 = 30 deg.
Relative inlet angle, β_1 = 60 deg.
Absolute outlet angle, α_2 = 60 deg.
Relative outlet angle, β_2 = 30 deg.

Determine,

(i) the axial velocity;
(ii) the mass flow rate;
(iii) the power absorbed by the stage;
(iv) the flow angles at the hub;
(v) the reaction ratio of the stage at the hub;

given that the rotational speed of the rotor is 6000 rev/min and the gas density is $1 \cdot 5$ kg/m^3 which can be assumed constant for the stage. It can be further assumed that stagnation enthalpy and entropy are constant before and after the rotor row for the purpose of simplifying the calculations.

(i) The rotational speed, $\Omega = 2\pi N/60 = 628 \cdot 4$ rad/s.

Therefore blade tip speed, $U_t = \Omega r_t = 314 \cdot 2$ m/s
and blade speed at hub, $U_h = \Omega r_h = 282 \cdot 5$ m/s.

From the velocity diagram for the stage (e.g. Fig. 5.2), the blade tip speed is

$$U_t = c_x (\tan 60° + \tan 30°) = c_x (\sqrt{3} + 1/\sqrt{3}).$$

Therefore $c_x = 136$ m/s, constant at all radii by eqn. (6.6a).

(ii) The rate of mass flow, $\dot{m} = \pi(r_t^2 - r_h^2)\rho c_x$

$$= \pi(0 \cdot 5^2 - 0 \cdot 45^2)1 \cdot 5 \times 136 = 30 \cdot 4 \text{ kg/s}. \quad \cdot$$

(iii) The power absorbed by the stage,

$$\begin{aligned}
\dot{W}_c &= \dot{m} \, U_t \, (c_{\theta 2t} - c_{\theta 1t}) \\
&= \dot{m} \, U_t \, c_x \, (\tan \alpha_{2t} - \tan \alpha_{1t}) \\
&= 30 \cdot 4 \times 314 \cdot 2 \times 136 (\sqrt{3} - 1/\sqrt{3}) \\
&= 1 \cdot 5 \text{ MW}.
\end{aligned}$$

(iv) At inlet to the rotor tip,

$$c_{\theta 1t} = c_x \tan \alpha_1 = 136/\sqrt{3} = 78 \cdot 6 \text{ m/s}.$$

The absolute flow is a free-vortex, $rc_\theta = $ constant.
Therefore $c_{\theta 1h} = c_{\theta 1t}(r_t/r_h) = 78 \cdot 6 \times 0 \cdot 5/0 \cdot 45 = 87 \cdot 3$ m/s.
At outlet to the rotor tip,

$$c_{\theta 2t} = c_x \tan \alpha_2 = 136 \times \sqrt{3} = 235 \cdot 6 \text{ m/s}.$$

Therefore $c_{\theta 2h} = c_{\theta 2t}(r_t/r_h) = 235 \cdot 6 \times 0 \cdot 5/0 \cdot 45 = 262$ m/s.
The flow angles at the hub are,

$$\begin{aligned}
\tan \alpha_1 &= c_{\theta 1h}/c_x = 87 \cdot 3/136 & &= 0 \cdot 642, \\
\tan \beta_1 &= U_h/c_x - \tan \alpha_1 & &= 1 \cdot 436, \\
\tan \alpha_2 &= c_{\theta 2h}/c_x = 262/136 & &= 1 \cdot 928, \\
\tan \beta_2 &= U_h/c_x - \tan \alpha_2 & &= 0 \cdot 152.
\end{aligned}$$

Thus $a_1 = 32.75°$, $\beta_1 = 55.15°$, $a_2 = 62.6°$, $\beta_2 = 8.64°$ at the hub.

(v) The reaction at the hub can be found by several methods. With eqn. (6.9)

$$R = 1 - k/r^2$$

and noticing that, from symmetry of the velocity triangles,

$$R = 0.5 \text{ at } r = r_t, \text{ then } k = 0.5 \, r_t{}^2.$$

Therefore $R_h = 1 - 0.5 \,(0.5/0.45)^2 = 0.382$.

The velocity triangles will be asymmetric and similar to those in Fig. 5.4(b).

The simplicity of the flow under free-vortex conditions is, superficially, very attractive to the designer and many compressors have been designed to conform to this flow. (Constant[1,2] may be consulted for an account of early British compressor design methods.) Figure 6.4 illustrates the variation of fluid angles and Mach numbers of a typical

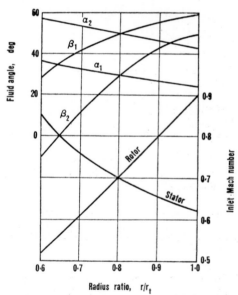

FIG. 6.4. Variation of fluid angles and Mach numbers of a free-vortex compressor stage with radius (adapted from Howell, Ref. 2 of Chapter 5).

compressor stage designed for free-vortex flow. Characteristic of this flow are the large fluid deflections near the inner wall and high Mach numbers near the outer wall, both effects being deleterious to efficient performance. A further serious disadvantage is the large amount of rotor twist from root to tip which adds to the expense of blade manufacture.

Many types of vortex design have been proposed to overcome some of the disadvantages set by free-vortex design and several of these are compared by Horlock.[3] Radial equilibrium solutions for the work and axial velocity distributions of some of these vortex flows in an axial compressor stage are given below.

2. Forced vortex

This is sometimes called "solid-body" rotation because c_θ varies directly with r. At entry to the rotor assume h_{01} is constant and $c_{\theta 1} = K_1 r$.

With eqn. (6.6a)

$$\frac{d}{dr}\left(\frac{c_{x1}^2}{2}\right) = -K_1 \frac{d}{dr}(K_1 r^2)$$

and, after integrating,

$$c_{x1}^2 = \text{constant} - 2K_1^2 r^2. \qquad (6.10)$$

After the rotor $c_{\theta 2} = K_2 r$ and $h_{02} - h_{01} = U(c_{\theta 2} - c_{\theta 1}) = \Omega(K_2 - K_1)r^2$. Thus, as the work distribution is non-uniform, the radial equilibrium equation in the form eqn. (6.6) is required for the flow after the rotor.

$$\frac{dh_{02}}{dr} = 2\Omega(K_2 - K_1)r = \frac{d}{dr}\left(\frac{c_{x2}^2}{2}\right) + K_2 \frac{d}{dr}(K_2 r^2).$$

After rearranging and integrating

$$c_{x2}^2 = \text{constant} - 2[K_2^2 - \Omega(K_2 - K_1)]r^2. \qquad (6.11)$$

The constants of integration in eqns. (6.10) and (6.11) can be found from the continuity of mass flow, i.e.

$$\frac{\dot{m}}{2\pi\rho} = \int_{r_h}^{r_t} c_{x1}r\,dr = \int_{r_h}^{r_t} c_{x2}r\,dr, \qquad (6.12)$$

which applies to the assumed incompressible flow.

3. *General whirl distribution*

The tangential velocity distribution is given by

$$c_{\theta 1} = ar^n - b/r \quad \text{(before rotor)}, \tag{6.13a}$$

$$c_{\theta 2} = ar^n + b/r \quad \text{(after rotor)}. \tag{6.13b}$$

The distribution of work for all values of the index n is constant with radius so that if h_{01} is uniform, h_{02} is also uniform with radius. From eqns. (6.13)

$$\Delta W = h_{02} - h_{01} = U(c_{\theta 2} - c_{\theta 1}) = 2b\Omega. \tag{6.14}$$

Selecting different values of n gives several of the tangential velocity distributions commonly used in compressor design. With $n = 0$, or zero power blading, it leads to the so-called "exponential" type of stage design (included as an exercise at the end of this chapter). With $n = 1$, or *first power blading*, the stage design is called (incorrectly, as it transpires later) "constant reaction".

First power stage design. For a given stage temperature rise the discussion in Chapter 5 would suggest the choice of 50% reaction at all radii for the highest stage efficiency. With swirl velocity distributions

$$c_{\theta 1} = ar - b/r, \qquad c_{\theta 2} = ar + b/r \tag{6.15}$$

before and after the rotor respectively, and rewriting the expression for reaction, eqn. (5.11), as

$$R = 1 - \frac{c_x}{2U}(\tan a_1 + \tan a_2), \tag{6.16}$$

then, using eqn. (6.15),

$$R = 1 - a/\Omega = \text{constant}. \tag{6.17}$$

Implicit in eqn. (6.16) is the assumption that the axial velocity across the rotor remains constant which, of course, is tantamount to ignoring radial equilibrium. The axial velocity *must* change in crossing the rotor row so that eqn. (6.17) is only a crude approximation at the best. Just how crude is this approximation will be indicated below.

Assuming constant stagnation enthalpy at entry to the stage, integrating eqn. (6.6a), the axial velocity distributions before and after the rotor are

$$c_{x1}^2 = \text{constant} - 4a(\tfrac{1}{2}ar^2 - b.\log r), \qquad (6.18a)$$

$$c_{x2}^2 = \text{constant} - 4a(\tfrac{1}{2}ar^2 + b.\log r). \qquad (6.18b)$$

More conveniently, these expressions can be written non-dimensionally as,

$$\left(\frac{c_{x1}}{U_t}\right)^2 = A_1 - \left(\frac{2a}{\Omega}\right)^2 \left[\frac{1}{2}\left(\frac{r}{r_t}\right)^2 - \frac{b}{ar_t^2}\log\left(\frac{r}{r_t}\right)\right], \qquad (6.19a)$$

$$\left(\frac{c_{x2}}{U_t}\right)^2 = A_2 - \left(\frac{2a}{\Omega}\right)^2 \left[\frac{1}{2}\left(\frac{r}{r_t}\right)^2 + \frac{b}{ar_t^2}\log\left(\frac{r}{r_t}\right)\right], \qquad (6.19b)$$

in which $U_t = \Omega r_t$ is the tip blade speed. The constants A_1, A_2 are not entirely arbitrary as the continuity equation, eqn. (6.12), must be satisfied.

EXAMPLE. As an illustration consider a single stage of an axial-flow air compressor of hub-tip ratio 0·4 with a nominally constant reaction (i.e. according to eqn. (6·17)) of 50%. Assuming incompressible, inviscid flow, a blade tip speed of 300 m/s, a blade tip diameter of 0·6 m, and a stagnation temperature rise of 16·1 °C, determine the radial equilibrium values of axial velocity before and after the rotor. The axial velocity far upstream of the rotor at the casing is 120 m/s. Take C_p for air as 1·005 kJ/(kg °C).

The constants in eqn. (6.19) can be easily determined. From eqn. (6.17)

$$2a/\Omega = 2(1 - R) = 1{\cdot}0.$$

Combining eqns. (6.14) and (6.17)

$$\frac{b}{ar_t^2} = \frac{\Delta W}{2\Omega^2(1 - R)r_t^2} = \frac{C_p.\Delta T_0}{2U_t^2(1 - R)}$$

$$= \frac{1005 \times 16{\cdot}1}{300^2} = 0{\cdot}18.$$

The inlet axial velocity distribution is completely specified and the constant A_1 solved. From eqn. (6.19a)

$$\left(\frac{c_{x1}}{U_t}\right)^2 = A_1 - [\tfrac{1}{2}(r/r_t)^2 - 0\cdot 18 \log(r/r_t)].$$

At $r = r_t$, $c_{x1}/U_t = 0\cdot 4$ and hence $A_1 = 0\cdot 66$.

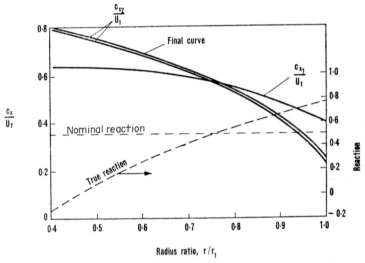

FIG. 6.5. Solution of exit axial-velocity profile for a first power stage.

Although an explicit solution for A_2 can be worked out from eqn. (6.19b) and eqn. (6.12), it is far speedier to use a semigraphical procedure. For an arbitrarily selected value of A_2, the distribution of c_{x2}/U_t is known. Values of $(r/r_t) \cdot (c_{x2}/U_t)$ and $(r/r_t) \cdot (c_{x1}/U_t)$ are plotted against r/r_t and the areas under these curves compared. New values of A_2 are then chosen until eqn. (6.12) is satisfied. This procedure is quite rapid and normally requires only two or three attempts to give a satisfactory solution. Figure 6.5 shows the final solution of c_{x2}/U_t obtained after three attempts. The solution is,

$$\left(\frac{c_{x2}}{U_t}\right)^2 = 0\cdot 56 - \left[\frac{1}{2}\left(\frac{r}{r_t}\right)^2 + 0\cdot 18 \log\left(\frac{r}{r_t}\right)\right].$$

It is illuminating to calculate the actual variation in reaction taking account of the change in axial velocity. From eqn. (5.10c) the true reaction across a normal stage is,

$$R' = \frac{w_1^2 - w_2^2}{2U(c_{\theta 2} - c_{\theta 1})}.$$

From the velocity triangles, Fig. 5.2,

$$w_1^2 - w_2^2 = (w_{\theta 1} + w_{\theta 2})(w_{\theta 1} - w_{\theta 2}) + (c_{x1}^2 - c_{x2}^2).$$

As $w_{\theta 1} + w_{\theta 2} = 2U - (c_{\theta 1} + c_{\theta 2})$ and $w_{\theta 1} - w_{\theta 2} = c_{\theta 2} - c_{\theta 1}$,

$$R' = 1 - \frac{c_{\theta 1} + c_{\theta 2}}{2U} + \frac{c_{x1}^2 - c_{x2}^2}{2U(c_{\theta 2} - c_{\theta 1})}.$$

For the first power swirl distribution, eqn. (6.15),

$$R' = 1 - \frac{a}{\Omega} + \frac{c_{x1}^2 - c_{x2}^2}{4\Omega b}.$$

From the radial equilibrium solution in eqns. (6.19), after some rearrangement,

$$\frac{c_{x1}^2 - c_{x2}^2}{4\Omega b} = \frac{A_1 - A_2}{2\psi_t} + \left(\frac{2a}{\Omega}\right) \log\left(\frac{r}{r_t}\right),$$

where

$$\psi_t = \frac{\Delta W}{U_t^2} = \frac{C_p \Delta T_0}{\Omega^2 r_t^2}.$$

In the above example, $1 - a/\Omega = \frac{1}{2}$, $\psi t = 0.18$

$$R' = 0.778 + \log(r/r_t).$$

The true reaction variation is shown in Fig. 6.5 and it is evident that eqn. (6.17) is *invalid* as a result of axial velocity changes.

THE DIRECT PROBLEM

The flow angle variation is specified in the direct problem and the radial equilibrium equation enables the solution of c_x and c_θ to be found. The general radial equilibrium equation can be written in the form

$$\frac{dh_0}{dr} - T\frac{ds}{dr} = \frac{c_\theta^2}{r} + c\frac{dc}{dr}$$

$$= \frac{c^2\sin^2 a}{r} + c\frac{dc}{dr}, \tag{6.20}$$

as $c_\theta = c\sin a$.

If both dh_0/dr and ds/dr are zero, eqn. (6.20) integrated gives

$$\log c = -\int \sin^2 a \frac{dr}{r} + \text{constant}$$

or, if $c = c_m$ at $r = r_m$, then

$$\frac{c}{c_m} = \exp\left(-\int_{r_m}^r \sin^2 a \frac{dr}{r}\right). \tag{6.21}$$

If the flow angle a is held constant, eqn. (6.21) simplifies still further,

$$\frac{c}{c_m} = \frac{c_x}{c_{xm}} = \frac{c_\theta}{c_{\theta m}} = \left(\frac{r}{r_m}\right)^{-\sin^2 a} \tag{6.22}$$

The vortex distribution represented by eqn. (6.22) is frequently employed in practice as untwisted blades are relatively simple to manufacture.

The general solution of eqn. (6.20) can be found by introducing a suitable *integrating factor* into the equation. Multiplying throughout by $\exp[2\int \sin^2 \alpha \, dr/r]$ it follows that

$$\frac{d}{dr}\left\{c^2 \exp[2\int\sin^2 \alpha \, dr/r]\right\} = 2\left(\frac{dh_0}{dr} - T\frac{ds}{dr}\right)\exp[2\int\sin^2 \alpha \, dr/r].$$

After integrating and inserting the limit $c = c_m$ at $r = r_m$, then

$$c^2 \exp\left[2\int^r \sin^2\alpha \, dr/r\right] - c_m^2 \exp\left[2\int^{r_m} \sin^2\alpha \, dr/r\right]$$

$$= 2\int_{r_m}^r \left(\frac{dh_0}{dr} - T\frac{ds}{dr}\right)\exp\left[2\int \sin^2\alpha \, dr/r\right]dr. \tag{6.23}$$

Particular solutions of eqn. (6.23) can be readily obtained for simple radial distributions of α, h_0 and s. Two solutions are considered here in which both $2dh_0/dr = kc_m^2/r_m$ and $ds/dr = 0$, k being an arbitrary constant

(i) Let $a = 2\sin^2\alpha$. Then $\exp[2\int\sin^2\alpha\, dr/r] = r^a$ and, hence

$$\left(\frac{c}{c_m}\right)^2\left(\frac{r}{r_m}\right)^a = 1 + \frac{k}{1+a}\left[\left(\frac{r}{r_m}\right)^{1+a} - 1\right]. \qquad (6.23a)$$

Equation (6.22) is obtained immediately from this result with $k = 0$.

(ii) Let $br/r_m = 2\sin^2\alpha$. Then,

$$c^2\exp(br/r_m) - c_m^2\exp(b) = (kc_m^2/r_m)\int_{r_m}^{r}\exp(br/r_m)dr$$

and eventually,

$$\left(\frac{c}{c_m}\right)^2 = \frac{k}{b} + \left(1 - \frac{k}{b}\right)\exp\left[b\left(1 - \frac{r}{r_m}\right)\right]. \qquad (6.23b)$$

COMPRESSIBLE FLOW THROUGH A FIXED BLADE ROW

In the blade rows of high-performance gas turbines, fluid velocities approaching, or even exceeding, the speed of sound are quite normal and compressibility effects may no longer be ignored. A simple analysis is outlined below for the inviscid flow of a perfect gas through a *fixed* row of blades which, nevertheless, can be extended to the flow through moving blade rows.

The radial equilibrium equation, eqn. (6.6), applies to *compressible* flow as well as incompressible flow. With constant stagnation enthalpy and constant entropy, a free-vortex flow therefore implies uniform axial velocity downstream of a blade row, regardless of any *density* changes incurred in passing through the blade row. In fact, for high-speed flows there *must* be a density change in the blade row which implies a streamline shift as shown in Fig. 6.1. This may be illustrated by considering the free-vortex flow of a perfect gas as follows. In radial equilibrium,

$$\frac{1}{\rho}\frac{dp}{dr} = \frac{c_\theta^2}{r} = \frac{K^2}{r^3} \quad \text{with } c_\theta = K/r.$$

For reversible adiabatic flow of a perfect gas $\rho = Ep^{1/\gamma}$, where E is constant. Thus

$$\int p^{-1/\gamma}dp = EK^2 \int r^{-3}dr + \text{constant},$$

therefore

$$p = \left[\text{constant} - \left(\frac{\gamma-1}{2\gamma} \right) \frac{EK^2}{r^2} \right]^{\gamma/(\gamma-1)}. \qquad (6.24)$$

For this free-vortex flow the pressure, and therefore the density also, must be larger at the casing than at the hub. The density difference from hub to tip may be appreciable in a high-velocity, high-swirl angle flow. If the fluid is without swirl at entry to the blades the density will be uniform. Therefore, from continuity of mass flow there must be a redistribution of fluid in its passage across the blade row to compensate for the changes in density. Thus, for this blade row, the continuity equation is,

$$\dot{m} = \rho_1 A_1 c_{x1} = 2\pi c_{x2} \int_{r_h}^{r_t} \rho_2 r dr, \qquad (6.25)$$

where ρ_2 is the density of the swirling flow, obtainable from eqn. (6.24).

CONSTANT SPECIFIC MASS FLOW

Although there appears to be no evidence that the redistribution of the flow across blade rows is a source of inefficiency, it has been suggested[4] that the radial distribution of c_θ for each blade row is chosen so that the product of axial velocity and density is constant with radius, i.e.

$$d\dot{m}/dA = \rho c_x = \rho c \cos \alpha = \rho_m c_m \cos \alpha_m = \text{constant} \qquad (6.26)$$

where subscript m denotes conditions at $r = r_m$.

This *constant specific mass flow design* is the logical choice when radial equilibrium theory is applied to compressible flows as the assumption that $c_r = 0$ is then likely to be realised.

Solutions may be determined by means of a simple numerical procedure and, as an illustration of one method, a turbine stage is considered here. It is convenient to assume that the stagnation enthalpy is uniform at nozzle entry, the entropy is constant throughout the stage and the fluid is a perfect gas. At nozzle exit under these conditions the equation of radial equilibrium, eqn. (6.20), can be written as

$$dc/c = -\sin^2 \alpha \, dr/r. \tag{6.27}$$

From eqn. (6.1), noting that at constant entropy the acoustic velocity $a = \sqrt{(dp/d\rho)}$,

$$\frac{1}{\rho} \frac{dp}{dr} = \frac{1}{\rho} \left(\frac{dp}{d\rho}\right)\left(\frac{d\rho}{dr}\right) = \frac{a^2}{\rho} \frac{d\rho}{dr} = \frac{c^2}{r} \sin^2 \alpha,$$

$$\therefore \; d\rho/\rho = M^2 \sin^2 \alpha \, dr/r \tag{6.28}$$

where the flow Mach number

$$M = c/a = c/\sqrt{(\gamma RT)}. \tag{6.28a}$$

The isentropic relation between temperature and density for a perfect gas is

$$T/T_m = (\rho/\rho_m)^{\gamma-1}$$

which after logarithmic differentiation gives

$$dT/T = (\gamma - 1)d\rho/\rho. \tag{6.29}$$

Using the above set of equations the procedure for determining the nozzle exit flow is as follows. Starting at $r = r_m$ values of c_m, α_m, T_m and ρ_m are assumed to be known. For a small finite interval Δr the changes in velocity Δc, density $\Delta\rho$, and temperature ΔT can be computed using eqns. (6.27), (6.28) and (6.29) respectively. Hence, at the new radius $r = r_m + \Delta r$ the velocity $c = c_m + \Delta c$, the density $\rho = \rho_m + \Delta\rho$ and temperature $T = T_m + \Delta T$ are obtained. The corresponding flow angle α and Mach number M can now be determined from eqns. (6.26) and (6.28a) respectively. Thus, all parameters of the problem are known at radius $r = r_m + \Delta r$. This procedure is repeated for further increments in radius to the casing and again from the mean radius to the hub.

Figure 6.6 shows the distributions of flow angle and Mach number computed with this procedure for a turbine nozzle blade row of 0·6 hub/tip radius ratio. The input data used was $\alpha_m = 70.4$ deg and

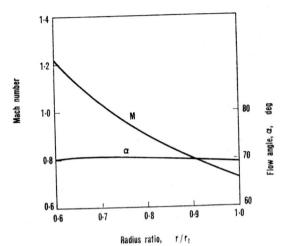

Fig. 6.6. Flow angle and Mach number distributions with radius of a nozzle blade row designed for constant specific mass flow.

$M = 0.907$ at the mean radius. Air was assumed at a stagnation pressure of 859 kPa and a stagnation temperature of 465 K. A remarkable feature of these results is the almost uniform swirl angle which is obtained.

With the nozzle exit flow fully determined the flow at rotor outlet can now be computed by a similar procedure. The procedure is a little more complicated than that for the nozzle row because the specific work done by the rotor is not uniform with radius. Across the rotor, using the notation of Chapter 4,

$$h_{o2} - h_{o3} = U(c_{\theta 2} + c_{\theta 3}) \qquad (6.30)$$

and hence the gradient in stagnation enthalpy after the rotor is

$$dh_{o3}/dr = - d[U(c_{\theta 2} + c_{\theta 3})]/dr = - d(Uc_{\theta 2})/dr - d(Uc_3 \sin \alpha_3)/dr.$$

After differentiating the last term,

$$-dh_o = d(Uc_{\theta 2}) + U(c \sin \alpha \, dr/r + \sin \alpha \, dc + c \cos \alpha \, d\alpha) \qquad (6.30a)$$

the subscript 3 having now been dropped.

From eqn. (6.20) the radial equilibrium equation applied to the rotor exit flow is

$$dh_o = c^2 \sin^2 \alpha \, dr/r + c \, dc. \qquad (6.30b)$$

After logarithmic differentiation of $\rho c \cos \alpha = $ constant,

$$d\rho/\rho + dc/c = \tan \alpha \, d\alpha. \tag{6.31}$$

Eliminating successively dh_o between eqns. (6.30a) and (6.30b), $d\rho/\rho$ between eqns. (6.28) and (6.31) and finally $d\alpha$ from the resulting equations gives

$$\frac{dc}{c}\left(1 + \frac{c_\theta}{U}\right) = -\sin^2\alpha \left\{\frac{d(rc_{\theta 2})}{rc_\theta} + \left(1 + \frac{c_\theta}{U} + M_x^2\right)\frac{dr}{r}\right\} \tag{6.32}$$

where $M_x = M \cos \alpha = c \cos \alpha / \sqrt{(\gamma RT)}$ and the static temperature

$$T = T_3 = T_{o3} - c_3^2/(2C_p)$$
$$= T_{o2} - [U(c_{\theta 2} + c_{\theta 3}) + \tfrac{1}{2}c_3^2]/C_p. \tag{6.33}$$

The verification of eqn. (6.32) is left as an exercise for the diligent student.

Provided that the exit flow angle α_3 at $r = r_m$ and the mean rotor blade speeds are specified, the velocity distribution, etc., at rotor exit can be readily computed from these equations.

OFF-DESIGN PERFORMANCE OF A STAGE

A turbine stage is considered here although, with some minor modifications, the analysis can be made applicable to a compressor stage.

Assuming the flow is at constant entropy, apply the radial equilibrium equation, eqn. (6.6), to the flow on both sides of the rotor, then

$$\frac{dh_{o3}}{dr} = \frac{dh_{o2}}{dr} - \Omega\frac{d}{dr}(rc_{\theta 2} + rc_{\theta 3}) = c_{x3}\frac{dc_{x3}}{dr} + \frac{c_{\theta 3}}{r}\frac{d}{dr}(rc_{\theta 3}).$$

Therefore

$$c_{x2}\frac{dc_{x2}}{dr} + \left(\frac{c_{\theta 2}}{r} - \Omega\right)\frac{d}{dr}(rc_{\theta 2}) = c_{x3}\frac{dc_{x3}}{dr} + \left(\frac{c_{\theta 3}}{r} + \Omega\right)\frac{d}{dr}(rc_{\theta 3}).$$

Substituting $c_{\theta 3} = c_{x3}\tan\beta_3 - \Omega r$ into the above equation, then, after some simplification,

$$c_{x2}\frac{dc_{x2}}{dr} + \left(\frac{c_{\theta 2}}{r} - \Omega\right)\frac{d}{dr}(rc_{\theta 2}) =$$

$$c_{x3}\frac{dc_{x3}}{dr} + \frac{c_{x3}}{r}\tan\beta_3\frac{d}{dr}(rc_{x3}\tan\beta_3) - 2\Omega c_{x3}\tan\beta_3. \tag{6.34}$$

In a particular problem the quantities $c_{x2}, c_{\theta2}, \beta_3$ are known functions of radius and Ω can be specified. Equation (6.34) is thus a first order differential equation in which c_{x3} is unknown and may best be solved, in the general case, by numerical iteration. This procedure requires a guessed value of c_{x3} at the hub and, by applying eqn. (6.34) to a small interval of radius Δr, a new value of c_{x3} at radius $r_h + \Delta r$ is found. By repeating this calculation for successive increments of radius a complete velocity profile c_{x3} can be determined. Using the continuity relation

$$\int_{r_h}^{r_t} c_{x3} r \, dr = \int_{r_h}^{r_t} c_{x2} r \, dr,$$

this initial velocity distribution can be integrated and a new, more accurate, estimate of c_{x3} at the hub then found. Using this value of c_{x3} the step-by-step procedure is repeated as described and again checked by continuity. This iterative process is normally rapidly convergent and, in most cases, three cycles of the calculation enables a sufficiently accurate exit velocity profile to be found.

The off-design performance may be obtained by making the approximation that the rotor relative exit angle β_3 and the nozzle exit angle α_2 remain constant at a particular radius with a change in mass flow. This approximation is not unrealistic as cascade data (see Chapter 3) suggest that fluid angles at outlet from a blade row alter very little with change in incidence up to the stall point.

Although any type of flow through a stage may be successfully treated using this method, rather more elegant solutions in closed form can be obtained for a few special cases. One such case is outlined below for a free-vortex turbine stage whilst other cases are already covered by eqns. (6.21)–(6.23).

FREE-VORTEX TURBINE STAGE

Suppose, for simplicity, a free-vortex stage is considered where, at the design point, the flow at rotor exit is completely axial (i.e. without swirl). At stage entry the flow is again supposed completely axial and of constant stagnation enthalpy h_{01}. Free-vortex conditions prevail at entry to the rotor, $rc_{\theta 2} = rc_{x2} \tan \alpha_2 = $ constant. The problem is to find how the axial velocity distribution at rotor exit varies as the mass flow is altered away from the design value.

At off-design conditions the relative rotor exit angle β_3 is assumed to remain equal to the value β^* at the design mass flow (* denotes design conditions). Thus, referring to the velocity triangles in Fig. 6.7, at off-design conditions the swirl velocity $c_{\theta 3}$ is evidently non-zero,

$$c_{\theta 3} = c_{x3} \tan \beta_3 - U$$

$$= c_{x3} \tan \beta_3^* - \Omega r. \tag{6.35}$$

At the design condition, $c_{\theta 3}^* = 0$ and so

$$c_{x3}^* \tan \beta_3^* = \Omega r. \tag{6.36}$$

Combining eqns. (6.35) and (6.36)

$$c_{\theta 3} = \Omega r \left(\frac{c_{x3}}{c_{x3}^*} - 1\right). \tag{6.37}$$

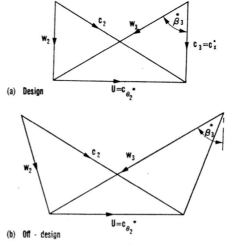

(a) **Design**

(b) **Off - design**

FIG. 6.7. Design and off-design velocity triangles for a free-vortex turbine stage.

The radial equilibrium equation at rotor outlet gives

$$\frac{dh_{03}}{dr} = c_{x3}\frac{dc_{x3}}{dr} + \frac{c_{\theta3}}{r}\frac{d}{dr}(rc_{\theta3}) = -\Omega\frac{d}{dr}(rc_{\theta3}), \qquad (6.38)$$

after combining with eqn. (6.33), noting that $dh_{02}/dr = 0$ and that $(d/dr)(rc_{\theta2}) = 0$ at all mass flows. From eqn. (6.37),

$$\Omega + \frac{c_{\theta3}}{r} = \Omega\frac{c_{x3}}{c_{x3}^*}, \qquad rc_{\theta3} = \Omega r^2\left(\frac{c_{x3}}{c_{x3}^*} - 1\right),$$

which when substituted into eqn. (6.38) gives,

$$-\frac{dc_{x3}}{dr} = \frac{\Omega^2}{c_{x3}^*}\left[2r\left(\frac{c_{x3}}{c_{x3}^*} - 1\right) + \frac{r^2}{c_{x3}^*}\frac{dc_{x3}}{dr}\right].$$

After rearranging,

$$\frac{dc_{x3}}{c_{x3} - c_{x3}^*} = \frac{-d(\Omega^2 r^2)}{(c_{x3}^{*2} + \Omega^2 r^2)}. \qquad (6.39)$$

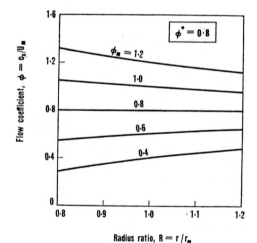

FIG. 6.8. Off-design rotor exit flow coefficients.

Equation (6.39) is immediately integrated in the form

$$\frac{c_{x3} - c_{x3}^*}{c_{x3m} - c_{x3}^*} = \frac{c_{x3}^{*2} + \Omega^2 r_m^2}{c_{x3}^{*2} + \Omega^2 r^2}, \tag{6.40}$$

where $c_{x3} = c_{x3m}$ at $r = r_m$. Equation (6.40) is more conveniently expressed in a non-dimensional form by introducing flow coefficients $\phi = c_{x3}/U_m$, $\phi^* = c_{x3}^*/U_m$ and $\phi_m = c_{x3m}/U_m$. Thus,

$$\frac{\phi/\phi^* - 1}{\phi_m/\phi^* - 1} = \frac{\phi^{*2} + 1}{\phi^{*2} + (r/r_m)^2}, \tag{6.40a}$$

If r_m is the mean radius then $c_{x3m} \doteqdot c_{x1}$ and, therefore, ϕ_m provides an approximate measure of the overall flow coefficient for the machine (N.B. c_{x1} is uniform).

The results of this analysis are shown in Fig. 6.8 for a representative design flow coefficient $\phi^* = 0.8$ at several different off-design flow coefficients ϕ_m, with $r/r_m = 0.8$ at the hub and $r/r_m = 1.2$ at the tip. It is apparent for values of $\phi_m < \phi^*$, that c_{x3} increases from hub to tip; conversely for $\phi_m > \phi^*$, c_{x3} decreases towards the tip.

The foregoing analysis is only a special case of the more general analysis of free-vortex turbine and compressor flows[5] in which rotor exit swirl, $rc_{\theta3}^*$ is constant (at design conditions), is included. However, from ref. 5, it is quite clear that even for fairly large values of a_{3m}^*, the value of ϕ is little different from the value found when $a_3^* = 0$, all other factors being equal. In Fig. 6.8 values of ϕ are shown when $a_{3m}^* = 31.4°$ at $\phi_m = 0.4$ ($\phi^* = 0.8$) for comparison with the results obtained when $a_3^* = 0$.

It should be noted that the rotor efflux flow at off-design conditions is *not* a free vortex.

ACTUATOR DISC APPROACH

In the radial equilibrium design method it was assumed that all radial motion took place within the blade row. However, in most turbomachines of low hub-tip ratio, appreciable radial velocities can be measured outside the blade row. Figure 6.9, taken from a review paper by Hawthorne and Horlock,[6] shows the distribution of the axial velocity component at various axial distances upstream and downstream of an isolated row of stationary inlet guide vanes. This figure clearly illustrates the appreciable redistribution of flow in regions outside of the blade row and that radial velocities must exist in these regions. For the flow through a single row of rotor blades, the variation in pressure (near the hub and tip) and variation in axial velocity (near the hub) both as functions of axial position, are shown in Fig. 6.10, also taken from Ref. 6. Clearly, radial equilibrium is not established entirely within the blade row.

A more accurate form of three-dimensional flow analysis than radial equilibrium theory is obtained with the *actuator disc* concept. The idea of an actuator disc is quite old and appears to have been first used in the theory of propellers; it has since evolved into a fairly sophisticated

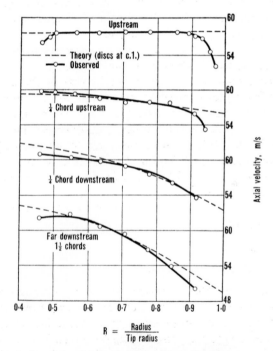

FIG. 6.9. Variation of the distribution in axial velocity through a row of guide vanes (adapted from Hawthorne and Horlock[6]).

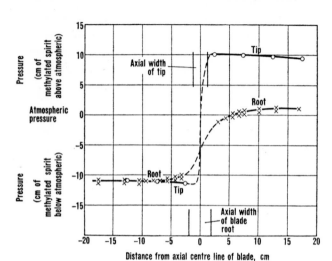

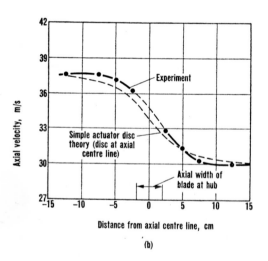

Fig. 6.10. (a) Pressure variation in the neighbourhood of a rotating blade row. (b) Axial velocity at the hub in the neighbourhood of a rotating blade row (adapted from Hawthorne and Horlock[6]).

method of analysing flow problems in turbomachinery. To appreciate the idea of an actuator disc, imagine that the axial width of each blade row is shrunk while, at the same time, the space–chord ratio, the blade angles and overall length of machine are maintained constant. As the deflection through each blade row for a given incidence is, apart from Reynolds number and Mach number effects (cf. Chapter 3 on cascades), fixed by the cascade geometry, a blade row of reduced width may be considered to affect the flow in exactly the same way as the original row. In the limit as the axial width vanishes, the blade row becomes, conceptually, a *plane discontinuity* of tangential velocity—the actuator disc. Note that while the tangential velocity undergoes an abrupt change in direction, the axial and radial velocities are continuous across the disc.

An isolated actuator disc is depicted in Fig. 6.11 with radial equili-

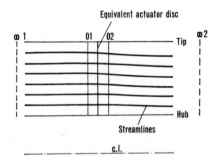

Equivalent actuator disc

Streamlines

c.l.

FIG. 6.11. The actuator disc assumption (after Horlock[3]).

brium established at fairly large axial distances from the disc. An approximate solution to the velocity fields upstream and downstream of the actuator can be found in terms of the axial velocity distributions, *far upstream* and *far downstream* of the disc. The detailed analysis exceeds the scope of this book, involving the solution of the equations of motion, the equation of continuity and the satisfaction of boundary conditions at the walls and disc. The form of the approximate solution is of considerable interest and is quoted below.

For convenience, conditions far upstream and far downstream of the disc are denoted by subscripts $\infty 1$ and $\infty 2$ respectively (Fig. 6.11). Actuator disc theory proves that at the disc ($x = 0$), at any given radius, the axial velocity is equal to the *mean* of the axial velocities at ∞1 and ∞2 at the *same* radius, or

$$c_{x01} = c_{x02} = \tfrac{1}{2}(c_{x\infty 1} + c_{x\infty 2}). \qquad (6.41)$$

Subscripts 01 and 02 denote positions immediately upstream and downstream respectively of the actuator disc. Equation (6.41) is known as the *mean-value rule*.

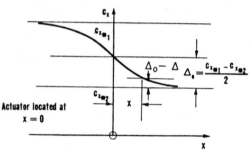

FIG. 6.12. Variation in axial velocity with axial distance from the actuator disc.

In the downstream flow field ($x \geq 0$), the *difference* in axial velocity at some position (x, r_A) to that at position ($x = \infty, r_A$) is conceived as a velocity perturbation. Referring to Fig. 6.12, the axial velocity pertubation at the disc ($x = 0, r_A$) is denoted by Δ_0 and at position (x, r_A) by Δ. The important result of actuator disc theory is that velocity perturbations *decay exponentially* away from the disc. This is also true for the upstream flow field ($x \leq 0$). The result obtained for the decay rate is

$$\Delta/\Delta_0 = 1 - \exp[\mp \pi x/(r_t - r_h)], \qquad (6.42)$$

where the minus and plus signs above apply to the flow regions $x \geq 0$ and $x \leq 0$ respectively. Equation (6.42) is often called the *settling-rate rule*. Since $c_{x1} = c_{x01} + \Delta$, $c_{x2} = c_{x02} - \Delta$ and noting that $\Delta_0 = \tfrac{1}{2}(c_{x\infty 1} - c_{x\infty 2})$, eqns. (6.41) and (6.42) combine to give,

$$c_{x1} = c_{x\infty 1} - \tfrac{1}{2}(c_{x\infty 1} - c_{x\infty 2})\exp[\pi x/(r_t - r_h)], \qquad (6.43a)$$

$$c_{x2} = c_{x\infty 2} + \tfrac{1}{2}(c_{x\infty 1} - c_{x\infty 2})\exp[-\pi x/(r_t - r_h)]. \qquad (6.43b)$$

At the disc, $x = 0$, eqns. (6.43) reduce to eqn. (6.41). It is of particular interest to note, in Figs. 6.9 and 6.10, how closely isolated actuator disc theory compares with experimentally derived results.

BLADE ROW INTERACTION EFFECTS

The spacing between consecutive blade rows in axial turbomachines is usually sufficiently small for mutual flow interactions to occur between the rows. This interference may be calculated by an extension of the results obtained from isolated actuator disc theory. As an illustration, the simplest case of two actuator discs situated a distance δ apart from one another is considered. The extension to the case of a large number of discs is given in ref. 6.

Consider each disc in turn as though it were in isolation. Referring to Fig. 6.13 disc A, located at $x = 0$, changes the far upstream velocity

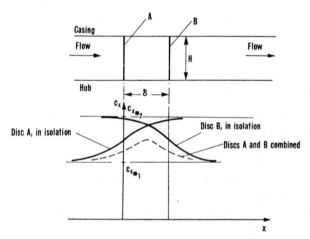

FIG. 6.13. Interaction between two closely spaced actuator discs.

$c_{x\infty 1}$ to $c_{x\infty 2}$ far downstream. Let us suppose for simplicity that the effect of disc B, located at $x = \delta$, exactly cancels the effect of disc A (i.e. the velocity far upstream of disc B is $c_{x\infty 2}$ which changes to $c_{x\infty 1}$ far downstream). Thus, for disc A in isolation,

$$c_x = c_{x\infty 1} - \tfrac{1}{2}(c_{x\infty 1} - c_{x\infty 2}) \exp\left[\frac{-\pi|x|}{H}\right], \quad x \leqq 0, \quad (6.44)$$

$$c_x = c_{x\infty 2} + \tfrac{1}{2}(c_{x\infty 1} - c_{x\infty 2}) \exp\left[\frac{-\pi|x|}{H}\right], \quad x \geqq 0, \quad (6.45)$$

where $|x|$ denotes modulus of x and $H = r_t - r_h$.

For disc B in isolation,

$$c_x = c_{x\infty 2} - \tfrac{1}{2}(c_{x\infty 2} - c_{x\infty 1}) \exp\left[\frac{-\pi|x - \delta|}{H}\right], \quad x \leq \delta, \qquad (6.46)$$

$$c_x = c_{x\infty 1} + \tfrac{1}{2}(c_{x\infty 2} - c_{x\infty 1}) \exp\left[\frac{-\pi|x - \delta|}{H}\right], \quad x \geq \delta. \qquad (6.47)$$

Now the combined effect of the two discs is most easily obtained by extracting from the above four equations the velocity perturbations appropriate to a given region and adding these to the related radial equilibrium velocity. For $x \leq 0$, add to $c_{x\infty 1}$ the perturbation velocities from eqns. (6.44) and (6.46).

$$c_x = c_{x\infty 1} - \tfrac{1}{2}(c_{x\infty 1} - c_{x\infty 2}) \left\{ \exp\left[\frac{-\pi|x|}{H}\right] - \exp\left[\frac{-\pi|x - \delta|}{H}\right] \right\}.$$

For the region $0 \leq x \leq \delta$, $\qquad\qquad\qquad\qquad\qquad\qquad (6.48)$

$$c_x = c_{x\infty 2} + \tfrac{1}{2}(c_{x\infty 1} - c_{x\infty 2}) \left\{ \exp\left[\frac{-\pi|x|}{H}\right] + \exp\left[\frac{-\pi|x - \delta|}{H}\right] \right\}.$$

For the region $x \geq \delta$, $\qquad\qquad\qquad\qquad\qquad\qquad\qquad (6.49)$

$$c_x = c_{x\infty 1} + \tfrac{1}{2}(c_{x\infty 1} - c_{x\infty 2}) \left\{ \exp\left[\frac{-\pi|x|}{H}\right] - \exp\left[\frac{-\pi|x - \delta|}{H}\right] \right\}.$$

$$(6.50)$$

Figure 6.13 indicates the variation of axial velocity when the two discs are regarded as *isolated* and when they are *combined*. It can be seen from the above equations that as the gap between these two discs is increased, so the perturbations tend to vanish. Thus in turbomachines where δ/r, is fairly small (e.g. the front stages of aircraft axial compressors or the rear stages of condensing steam turbines), interference effects are strong and one can infer that the simpler radial equilibrium analysis is then inadequate.

COMPUTER-AIDED METHODS OF SOLVING THE THROUGH-FLOW PROBLEM

Although actuator disc theory has given a better understanding of the complicated meridional (the radial–axial plane) through-flow problem in turbomachines of simple geometry and flow conditions, its application to the design of axial-flow compressors has been rather limited. The extensions of actuator disc theory to the solution of the complex three-dimensional, compressible flows in compressors with varying hub and tip radii and non-uniform total pressure distributions were found to have become too unwieldy in practice. In recent years advanced computational methods have been successfully evolved for predicting the meridional compressible flow in turbomachines with flared annulus walls. Gostelow *et al.*[11] have reviewed the recent developments in the case of axial-flow compressor design and they include an outline of the two most widely used techniques for solving the through-flow problem. The two different methods are:

  (i) streamline curvature[4,12];
  (ii) matrix through-flow analysis[13].

Both methods solve the same equations of fluid motion, energy and state for an axisymmetric flow through a turbomachine with varying hub and tip radii and therefore lead to the same solution. In the first method the equation for the meridional velocity $c_m = (c_r^2 + c_x^2)^{\frac{1}{2}}$ in a plane (at $x = x_a$) contain terms involving both the slope and curvature of the meridional streamlines which are estimated by using a polynomial curve-fitting procedure through points of equal stream function on neighbouring planes at $(x_a - \mathrm{d}x)$ and $(x_a + \mathrm{d}x)$. The major source of difficulty is in accurately estimating the curvature of the streamlines. In the second method a grid of calculating points is formed on which the stream function is expressed as a quasi-linear equation. A set of corresponding finite difference equations are formed which are then solved at all mesh points of the grid. A more detailed description of these methods is rather beyond the scope of the present text.

SECONDARY FLOWS

No account of three-dimensional motion in axial turbomachines would be complete without giving, at least, a brief description of secondary flow. When a fluid particle possessing rotation is turned (e.g. by a cascade) its axis of rotation is deflected in a manner analogous to the motion of a gyroscope, i.e. in a direction perpendicular to the direction of turning. The result of turning the rotation (or vorticity) vector is the formation of *secondary flows*. The phenomenon must occur to some degree in all turbomachines but is particularly in evidence in axial-flow compressors because of the thick boundary layers on the annulus walls. This case has been discussed in some detail by Horlock,[4] Preston,[7] Carter[8] and many other writers.

Consider the flow at inlet to the guide vanes of a compressor to be completely axial and with a velocity profile as illustrated in Fig. 6.14.

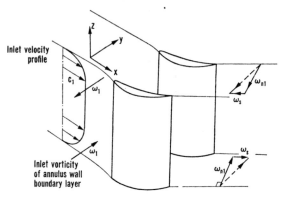

Fig. 6.14. Secondary vorticity produced by a row of guide vanes.

This velocity profile is non-uniform as a result of friction between the fluid and the wall; the vorticity of this boundary layer is normal to the approach velocity c_1 and of magnitude

$$\omega_1 = \frac{dc_1}{dz},\qquad(6.51)$$

where z is distance from the wall.

The direction of ω_1 follows from the right-hand screw rule and it will be observed that ω_1 is in opposite directions on the two annulus walls. This vector is turned by the cascade, thereby generating *secondary vorticity* parallel to the outlet stream direction. If the deflection angle

ϵ is not large, the magnitude of the secondary vorticity ω_s is, approximately,

$$\omega_s = -2\epsilon \frac{dc_1}{dz}. \tag{6.52}$$

A swirling motion of the cascade exit flow is associated with the vorticity ω_s, as shown in Fig. 6.15, which is in opposite directions for the two wall boundary layers. This secondary flow will be the *integrated* effect of the distribution of secondary vorticity along the blade length.

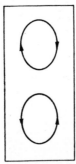

FIG. 6.15. Secondary flows at exit from a blade passage (viewed in upstream direction).

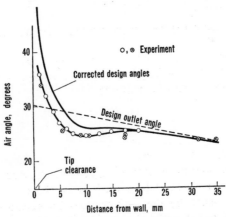

FIG. 6.16. Exit air angle from inlet guide vanes (adapted from Horlock[10]).

Now if the variation of c_1 with z is known or can be predicted, then the distribution of ω_s along the blade can be found using eqn. (6.52). By considering the secondary flow to be a small perturbation of the two-dimensional flow from the vanes, the flow angle distribution can be calculated using a series solution developed by Hawthorne.[9] The actual analysis lies outside the scope (and purpose) of this book, however. Experiments on cascades show excellent agreement with these calculations provided there are but small viscous effects and no flow separations. Such a comparison has been given by Horlock[10] and a typical result is shown in Fig. 6.16. It is clear that the flow is *overturned* near the walls and *underturned* some distance away from the walls. It is known that this overturning is a source of inefficiency in compressors as it promotes stalling at the blade extremities.

REFERENCES

1. CONSTANT, H., The early history of the axial type of gas turbine engine. *Proc. Instn. Mech. Engrs. London,* **153** (1945). (War emergency issues).
2. CONSTANT, H., *Gas Turbines and their Problems.* Todd, London (1953).
3. HORLOCK, J. H., *Axial Flow Compressors.* Butterworths, London (1958).
4. HORLOCK, J. H., *Axial Flow Turbines.* Butterworths, London (1966).
5. HORLOCK, J. H. and DIXON, S. L., The off-design performance of free vortex turbine and compressor stages. University of Liverpool, Dept. of Mech. Eng. Report ULME 66 (1965). Now A.R.C. 27, 612 (1966).
6. HAWTHORNE, W. R. and HORLOCK, J. H., Actuator disc theory of the incompressible flow in axial compressors. *Proc. Instn. Mech. Engrs. London,* **176** (1962).
7. PRESTON, J. H., A simple approach to the theory of secondary flows. *Aero. Quart.* **5** (Part 3) (1953).
8. CARTER, A. D. S., Three-dimensional flow theories for axial compressors and turbines. *Proc. Instn. Mech. Engrs. London,* **159** (1948).
9. HAWTHORNE, W. R., Some formulae for the calculation of secondary flow in cascades. Aero. Res. Comm. Report No. 17,519 (1955).
10. HORLOCK, J. H., Annulus wall boundary layers in axial compressor stages. *Trans. Am. Soc. Mech. Engrs.* Series D, **85** (1963).
11. GOSTELOW, J. P., HORLOCK, J. H. and MARSH, H., Recent developments in the aerodynamic design of axial flow compressors. Symposium at Warwick University. *Proc. Instn. Mech. Engrs.* **183**, Pt. 3N (1969).
12. SMITH, L. H., Jr., The radial-equilibrium equation of turbomachinery. *Trans. Am. Soc. Mech. Engrs.,* Series A, **88** (1966).
13. MARSH, H., A digital computer program for the through-flow fluid mechanics on an arbitrary turbomachine using a matrix method. *Aero. Res. Coun. Rep. Mem.* 3509 (1968).

PROBLEMS

1. Derive the radial equilibrium equation for an incompressible fluid flowing with axisymmetric swirl through an annular duct.

Air leaves the inlet guide vanes of an axial flow compressor in radial equilibrium and with a free-vortex tangential velocity distribution. The absolute static pressure and static temperature at the hub, radius 0·3 m, are 94·5 kPa and 293 K respectively. At the casing, radius 0·4 m, the absolute static pressure is 96·5 kPa. Calculate the flow angles at exit from the vanes at the hub and casing when the inlet absolute stagnation pressure is 101·3 kPa. Assume the fluid to be inviscid and incompressible. (Take $R = 0.287$ kJ/(kg °C) for air.)

2. A gas turbine stage has an initial absolute pressure of 350 kPa and a temperature of 565°C with negligible initial velocity. At the mean radius, 0·36 m, conditions are as follows:

Nozzle exit flow angle	68 deg
Nozzle exit absolute static pressure	207 kPa
Stage reaction	0·2

Determine the flow coefficient and stage loading factor at the mean radius and the reaction at the hub, radius 0·31 m, at the design speed of 8000 rev/min, given that the stage is to have a free vortex swirl at this speed. You may assume that losses are absent. Comment upon the results you obtain.

(Take $C_p = 1.148$ kJ/(kg °C) and $\gamma = 1.33$.)

3. Gas enters the nozzles of an axial flow turbine stage with uniform total pressure at a uniform velocity c_1 in the axial direction and leaves the nozzles at a constant flow angle a_2 to the axial direction. The absolute flow leaving the rotor c_3 is completely axial at all radii.

Using radial equilibrium theory and assuming no losses in total pressure show that

$$(c_3{}^2 - c_1{}^2)/2 = U_m\, c_{\theta m2} \left[1 - \left(\frac{r}{r_m}\right)^{\cos^2 a_2}\right]$$

where U_m is the mean blade speed,

$c_{\theta m2}$ is the tangential velocity component at nozzle exit at the mean radius $r = r_m$.

(*Note*: The approximation $c_3 = c_1$ at $r = r_m$ is used to derive the above expression.)

4. Gas leaves an untwisted turbine nozzle at an angle a to the axial direction and in radial equilibrium. Show that the variation in axial velocity from root to tip, assuming total pressure is constant, is given by

$$c_x r^{\sin^2 a} = \text{constant}.$$

Determine the axial velocity at a radius of 0·6 m when the axial velocity is 100 m/s at a radius of 0·3 m. The outlet angle a is 45 deg.

5. The flow at the entrance and exit of an axial flow compressor rotor is in radial equilibrium. The distributions of the tangential components of absolute velocity with radius are:

$$c_{\theta 1} = ar - b/r, \text{ before the rotor,}$$

$$c_{\theta 2} = ar + b/r, \text{ after the rotor,}$$

where a and b are constants. What is the variation of work done with radius? Deduce expressions for the axial velocity distributions before and after the rotor, assuming incompressible flow theory and that the radial gradient of stagnation pressure is zero.

At the mean radius, $r = 0.3$ m, the stage loading coefficient, $\psi = \Delta W/U_t^2$ is 0.3, the reaction ratio is 0.5 and the mean axial velocity is 150 m/s. The rotor speed is 7640 rev/min. Determine the rotor flow inlet and outlet angles at a radius of 0.24 m given that the hub/tip ratio is 0.5. Assume that at the mean radius the axial velocity remained unchanged ($c_{x1} = c_{x2}$ at $r = 0.3$ m).

(*Note*: ΔW is the specific work and U_t the blade tip speed.)

6. An axial flow turbine stage is to be designed for free-vortex conditions at exit from the nozzle row and for zero swirl at exit from the rotor. The gas entering the stage has a stagnation temperature of 1000 K, the mass flow rate is 32 kg/s, the root and tip diameters are 0.56 m and 0.76 m respectively, and the rotor speed is 8000 rev/min. At the rotor tip the stage reaction is 50% and the axial velocity is constant at 183 m/s. The velocity of the gas entering the stage is equal to that leaving.
Determine:

  (i)   the maximum velocity leaving the nozzles;
  (ii) the maximum absolute Mach number in the stage;
  (iii) the root section reaction;
  (iv) the power output of the stage;
  (v)   the stagnation and static temperatures at stage exit.

(Take $R = 0.287$ kJ/(kg °C) and $Cp = 1.147$ kJ/(kg °C).)

7. The rotor blades of an axial flow turbine stage are 100 mm long and are designed to receive gas at an incidence of 3 deg from a nozzle row. A free-vortex whirl distribution is to be maintained between nozzle exit and rotor entry. At rotor exit the absolute velocity is 150 m/s in the axial direction at all radii. The deviation is 5 deg for the rotor blades and zero for the nozzle blades at all radii. At the hub, radius 200 mm, the conditions are as follows:

Nozzle outlet angle	70 deg
Rotor blade speed	180 m/s
Gas speed at nozzle exit	450 m/s

Assuming that the axial velocity of the gas is constant across the stage, determine

  (i)   the nozzle outlet angle at the tip;
  (ii) the rotor blade inlet angles at hub and tip;
  (iii) the rotor blade outlet angles at hub and tip;
  (iv) the degree of reaction at root and tip.

Why is it essential to have a *positive* reaction in a turbine stage?

8. The rotor and stator of an isolated stage in an axial-flow turbomachine are to be represented by two actuator discs located at axial positions $x = 0$ and $x = \delta$ respectively. The hub and tip diameters are constant and the hub/tip radius ratio r_h/r_t is 0.5. The rotor disc considered on its own has an axial velocity of 100 m/s far upstream and 150 m/s downstream at a constant radius $r = 0.75\ r_t$. The stator disc

in isolation has an axial velocity of 150 m/s far upstream and 100 m/s far downstream at radius $r = 0.75\ r_t$. Calculate and plot the axial velocity variation between $-0.5 \leqslant x/r_t \leqslant 0.6$ at the given radius for each actuator disc in isolation and for the combined discs when

$$\text{(i) } \delta = 0.1\ r_t, \text{ (ii) } \delta = 0.25\ r_t, \text{ (iii) } \delta = r_t.$$

CHAPTER 7

Centrifugal Pumps, Fans and Compressors

And to thy speed add wings. (MILTON, *Paradise Lost.*)

INTRODUCTION

This chapter is concerned with the elementary fluid behaviour of *radial-flow* work-absorbing turbomachines comprising pumps, fans and compressors. The major part of the discussion is centred around the compressor since the basic action of all these machines is, in most respects, the same.

Turbomachines employing centrifugal effects for increasing fluid pressure have been in use for more than a century. The earliest machines using this principle were, undoubtedly, hydraulic pumps followed later by ventilating fans and blowers. It is on record[1] that a centrifugal compressor was used in one of the first aircraft jet propulsion engines. Development of the centrifugal compressor continued into the mid-1950s in this field, but long before this it had become abundantly clear [2,3] that for the increasingly larger engines required for aircraft propulsion the axial flow compressor was preferred. Not only was the frontal area (and drag) smaller with engines using axial compressors but also the efficiency for the same duty was better by as much as 3 or 4%. However, at very low air mass flow rates the efficiency of axial compressors drops sharply, blading is small and difficult to make accurately and the advantage lies with the centrifugal compressor.

In the mid-1960s the need for advanced military helicopters powered by small gas turbine engines provided the necessary impetus for further rapid development of the centrifugal compressor. The technological advances made in this sphere provided a spur to designers in a much wider field of existing centrifugal compressor applications, e.g. in small gas turbines for road vehicles and commercial helicopters as well as for diesel engine turbochargers, chemical plant processes, factory workshop air supplies and large-scale air-conditioning plant, etc. Recent performance data for small single-stage centrifugal compressors[4] quotes total to static efficiencies of 80–84% for pressure ratios between 4 and 6 to 1. Higher pressure ratios than this have been achieved in single stages but at reduced efficiency and small airflow range, e.g. Schorr *et al.*[12] designed and tested a centrifugal compressor which gave a pressure ratio of 10 to 1 at an efficiency of 72% but with an airflow range of only 10% at the design rotational speed.

SOME DEFINITIONS

Most of the pressure-increasing turbomachines in use are of the radial-flow type and vary from fans that produce pressure rises equivalent to a few millimetres of water to pumps producing heads of many hundreds of metres of water. The term *pump* is used when referring to machines that increase the pressure of a flowing liquid. The term *fan* is used for machines imparting only a small increase in pressure to a flowing gas. In this case the pressure rise is usually so small that the gas can be considered as being incompressible. A *compressor* gives a substantial rise in pressure to a flowing gas. For purposes of definition, the boundary between fans and compressors is often taken as that where the density ratio across the machine is 1·05. Sometimes, but more rarely nowadays, the term *blower* is used instead of fan.

A centrifugal compressor or pump consists essentially of a rotating *impeller* followed by *diffuser*. Figure 7.1 shows diagrammatically the various elements of a centrifugal compressor. Fluid is drawn in through the *inlet casing* into the *eye* of the impeller. The function of the impeller is to increase the energy level of the fluid by whirling it outwards, thereby increasing the angular momentum of the fluid. Both the static pressure and the velocity are increased within the impeller. The purpose of the diffuser is to convert the kinetic energy of the fluid leaving the impeller into pressure energy. This process can be accomplished by free diffusion in the annular space surrounding the impeller or, as indicated in Fig. 7.1, by incorporating a row of fixed diffuser vanes

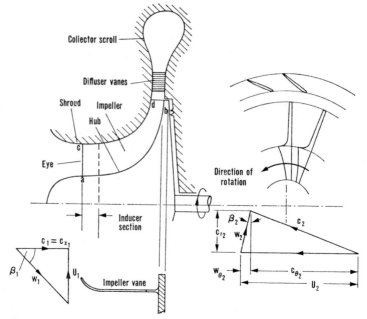

FIG. 7.1. Centrifugal compressor stage and velocity diagrams at impeller entry
and exit.

which allows the diffuser to be made very much smaller. Outside the
diffuser is a *scroll* or *volute* whose function is to collect the flow from
the diffuser and deliver it to the outlet pipe. Often, in low-speed compressors and pump applications where simplicity and low cost count
for more than efficiency, the volute follows immediately after the
impeller.

The *hub* is the curved surface of revolution of the impeller *a–b*; the
shroud is the curved surface *c–d* forming the outer boundary to the
flow of fluid. Impellers may be enclosed by having the shroud attached
to the vane ends (called shrouded impellers) or unenclosed with a small
clearance gap between the vane ends and the stationary wall. Whether
or not the impeller is enclosed the surface, *c–d* is generally called the
shroud. Shrouding an impeller has the merit of eliminating tip leakage
losses but at the same time increases friction losses. NACA tests have

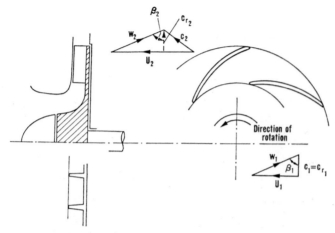

FIG. 7.2. Radial-flow pump and velocity triangles.

demonstrated that shrouding of a single impeller appears to be detrimental at high speeds and beneficial at low speeds. At entry to the
impeller the relative flow has a velocity w_1 at angle β_1 to the axis of
rotation. This relative flow is turned into the axial direction by the
inducer section or *rotating guide vanes* as they are sometimes called.
The inducer starts at the eye and usually finishes in the region where the
flow is beginning to turn into the radial direction. Some compressors
of advanced design extend the inducer well into the radial flow region
apparently to reduce the amount of relative diffusion.

To simplify manufacture and reduce cost, many fans and pumps are
confined to a two-dimensional radial section as shown in Fig. 7.2.
With this arrangement some loss in efficiency can be expected. For the
purpose of greatest utility, relations obtained in this chapter are generally
in terms of the three-dimensional compressor configuration.

THEORETICAL ANALYSIS OF A CENTRIFUGAL COMPRESSOR

The flow through a compressor stage is a highly complicated, three-
dimensional motion and a full analysis presents many problems of the
highest order of difficulty. However, we can obtain approximate

solutions quite readily by simplifying the flow model. We adopt the so-called *one-dimensional* approach which assumes that the fluid conditions are uniform over certain flow cross-sections. These cross-sections are conveniently taken immediately before and after the impeller as well as at inlet and exit of the entire machine. Where inlet vanes are used to give prerotation to the fluid entering the impeller, the one-dimensional treatment is no longer valid and an extension of the analysis is then required (see Chapter 6).

INLET CASING

The fluid is accelerated from velocity c_0 to velocity c_1 and the static pressure falls from p_0 to p_1 as indicated in Fig. 7.3. Since the stagnation enthalpy is constant in steady, adiabatic flow without shaft work then $h_{00} = h_{01}$ or,

$$h_0 + \tfrac{1}{2}c_0^2 = h_1 + \tfrac{1}{2}c_1^2.$$

Some efficiency definitions appropriate to this process are stated in Chapter 2.

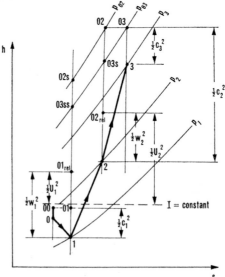

FIG. 7.3. Mollier diagram for the complete centrifugal compressor stage.

IMPELLER

The specific work done on the fluid is equal to the stagnation enthalpy rise and is also expressed by Euler's pump equation, eqn. (2.12a). Thus,

$$\Delta W = \frac{\dot{W}_c}{\dot{m}} = U_2 c_{\theta 2} - U_1 c_{\theta 1} = h_{02} - h_{01}. \tag{7.1}$$

Rearranging the above equation and putting $h_0 = h + \tfrac{1}{2}c^2$, then

$$h_1 + \tfrac{1}{2}c_1{}^2 - U_1 c_{\theta 1} = h_2 + \tfrac{1}{2}c_2{}^2 - U_2 c_{\theta 2} = I.$$

The function I is a fundamental property of some importance in the study of relative rotating flows involving radial motion and has lately acquired the name *rothalpy*, a contraction of rotational stagnation enthalpy. The rothalpy has the same value at impeller inlet and outlet so that it must be constant along flow lines between these stations. This can be written generally as,

$$I = h + \tfrac{1}{2}c^2 - Uc_\theta. \tag{7.2}$$

The general three-dimensional motion has components of velocity c_r, c_θ, and c_x respectively in the radial, tangential and axial directions and $c^2 = c_r^2 + c_\theta^2 + c_x^2$.

Thus,

$$I = h + \tfrac{1}{2}(c_r^2 + c_\theta^2 + c_x^2 - 2Uc_\theta).$$

Adding and subtracting $\tfrac{1}{2}U^2$ this becomes

$$I = h + \tfrac{1}{2}\{(U - c_\theta)^2 + c_r^2 + c_x^2 - U^2\}. \tag{7.2a}$$

From the velocity triangle, Fig. 7.1, $U - c_\theta = w_\theta$ and together with $w^2 = c_r^2 + w_\theta^2 + c_x^2$, eqn. (7.2a) becomes

$$I = h + \tfrac{1}{2}(w^2 - U^2)$$

or

$$I = h_{0\ rel} - \tfrac{1}{2}U^2,$$

since

$$h_{0\ rel} = h + \tfrac{1}{2}w^2. \tag{7.3}$$

Since $I_1 = I_2$ across the impeller, then eqn. (7.3) gives

$$h_2 - h_1 = \tfrac{1}{2}(U_2^2 - U_1^2) + \tfrac{1}{2}(w_1^2 - w_2^2). \tag{7.4}$$

The above expression provides the reason why the static enthalpy rise in a centrifugal compressor is so large compared with a single-stage axial compressor. On the right-hand side of eqn. (7.4), the second term $\frac{1}{2}(w_2^2 - w_1^2)$, is the contribution from the diffusion of relative velocity and was obtained for axial compressors also. The first term, $\frac{1}{2}(U_2^2 - U_1^2)$, is the contribution due to the centrifugal action which is zero if the streamlines remain at the same radii before and after the impeller.

The relation between state points 1 and 2 in Fig. 7.3 can be easily traced with the aid of eqn. (7.4).

Referring to Fig. 7.1, and in particular the inlet velocity diagram, the absolute flow has no whirl component or angular momentum and $c_{\theta 1} = 0$. In centrifugal compressors and pumps this is the normal situation where the flow is free to enter axially. For such a flow the specific work done on the fluid, from eqn. (7.1), is written as

$$\Delta W = U_2 c_{\theta 2} = h_{02} - h_{01} \qquad (7.1a)$$

in the case of compressors, and as

$$\Delta W = U_2 c_{\theta 2} = gH_i \qquad (7.1b)$$

in the case of pumps, where H_i (the "ideal" head) is the total head rise across the pump excluding all internal losses. In exceptional cases it may be necessary to impart *prerotation* to the flow entering the impeller as a means of reducing a high relative inlet velocity. The effects of high relative velocity at the impeller inlet are felt as Mach number effects in compressors and cavitation effects in pumps. The usual method of establishing prerotation requires the installation of a row of inlet guide vanes somewhere upstream of the impeller, the location depending upon the type of inlet. Unless contrary statements are made it will be assumed for the remainder of this chapter that there is no prerotation (i.e. $c_{\theta 1} = 0$).

DIFFUSER

The fluid is decelerated adiabatically from velocity c_2 to a velocity c_3, the static pressure rising from p_2 to p_3 as shown in Fig. 7.3. As the volute and outlet diffuser involve some further deceleration it is convenient to group the whole diffusion together as the change of state from point 2 to point 3. As the stagnation enthalpy in steady adiabatic flow without shaft work is constant, $h_{02} = h_{03}$ or $h_2+\frac{1}{2}c_2^2 = h_3+\frac{1}{2}c_3^2$. The process 2 to 3 in Fig. 7.3 is drawn as irreversible, there being a loss in stagnation pressure $p_{02}-p_{03}$ during the process.

INLET VELOCITY LIMITATIONS

The inlet eye is an important critical region in centrifugal pumps and compressors requiring careful consideration at the design stage. If the relative velocity of the inlet flow is too large in pumps, cavitation may result with consequent blade erosion or even reduced performance. In compressors large relative velocities can cause an increase in the impeller total pressure losses. In high-speed centrifugal compressors Mach number effects may become important with high relative velocities in the inlet. By suitable sizing of the eye the maximum relative velocity, or some related parameter, can be minimised to give the optimum inlet flow conditions. As an illustration the following analysis shows a simple optimisation procedure for a low-speed compressor based upon incompressible flow theory.

For the inlet geometry shown in Fig. 7.1, the absolute eye velocity is assumed to be uniform and axial. The inlet relative velocity is $w_1 = (c_{x1}^2 + U^2)^{\frac{1}{2}}$ which is clearly a maximum at the inducer tip radius r_{s1}. The volume flow rate is

$$Q = c_{x1}A_1 = \pi(r_{s1}^2 - r_{h1}^2)\,(w_{s1}^2 - \Omega^2 r_{s1}^2)^{\frac{1}{2}}. \tag{7.5}$$

It is worth noticing that with both Q and r_{h1} fixed:

(i) if r_{s1} is made large then, from continuity, the axial velocity is low but the blade speed is high,

(ii) if r_{s1} is made small the blade speed is small but the axial velocity is high.

Both extremes produce large relative velocities and there must exist some optimum radius r_{s1} for which the relative velocity is a minimum.

For maximum volume flow, differentiate eqn. (7.5) with respect to r_{s1} (keeping w_{s1} constant) and equate to zero,

$$\frac{1}{\pi}\frac{\partial Q}{\partial r_{s1}} = 0 = 2r_{s1}(w_{s1}^2 - \Omega^2 r_{s1}^2)^{\frac{1}{2}} - (r_{s1}^2 - r_{h1}^2)\,\Omega^2 r_{s1}/(w_{s1}^2 - \Omega^2 r_{s1}^2)^{\frac{1}{2}}$$

After simplifying,

$$2(w_{s1}^2 - \Omega^2 r_{s1}^2) = (r_{s1}^2 - r_{h1}^2)\Omega^2,$$
$$\therefore 2c_{x1}^2 = kU_{s1}^2,$$

where $k = 1 - (r_{h1}/r_{s1})^2$ and $U_{s1} = \Omega r_{s1}$. Hence, the optimum inlet velocity coefficient is

$$\phi = c_{x1}/U_{s1} = \cot \beta_{s1} = (k/2)^{\frac{1}{2}}. \tag{7.6}$$

Equation (7.6) specifies the optimum conditions for the inlet velocity triangles in terms of the hub/tip radius ratio. For typical values of this ratio (i.e. $0.3 \leq r_{h1}/r_{s1} \leq 0.6$) the optimum relative flow angle at the inducer tip β_{s1} lies between 56 deg and 60 deg.

OPTIMUM DESIGN OF A PUMP INLET

As discussed in Chapter 1, cavitation commences in a flowing liquid when the decreasing local static pressure becomes approximately equal to the vapour pressure, p_v. To be exact it is necessary to assume that gas cavitation is negligible and that sufficient nuclei exist in the liquid to initiate vapour cavitation.

The pump considered in the following analysis is again assumed to have the flow geometry shown in Fig. 7.1. Immediately upstream of the impeller blades the static pressure is $p_1 = p_{01} - \frac{1}{2}\rho c_{x1}^2$ where p_{01} is the stagnation pressure and c_{x1} is the axial velocity. In the vicinity of the impeller blades leading edges on the suction surfaces there is normally a rapid velocity increase which produces a further decrease in pressure. At cavitation inception the dynamic action of the blades causes the *local* pressure to reduce such that $p = p_v = p_1 - \sigma_b(\frac{1}{2}\rho w_1^2)$. The parameter σ_b which is the *blade cavitation coefficient* corresponding to the cavitation inception point, depends upon the blade shape and the flow incidence angle. For conventional pumps[11] operating normally this coefficient lies in the range $0.2 \leq \sigma_b \leq 0.4$. Thus, at cavitation inception,

$$p_1 = p_{01} - \tfrac{1}{2}\rho c_{x1}^2 = p_v + \sigma_b(\tfrac{1}{2}\rho w_1^2)$$
$$\therefore gH_s = (p_{01} - p_v)/\rho = \tfrac{1}{2}c_{x1}^2 + \sigma_b(\tfrac{1}{2}w_1^2) = \tfrac{1}{2}c_{x1}^2(1 + \sigma_b) + \tfrac{1}{2}\sigma_b U_{s1}^2$$

where H_s is the net positive suction head introduced earlier and it is implied that this is measured at the shroud radius $r = r_{s1}$.

To obtain the optimum inlet design conditions consider the suction specific speed defined as $\Omega_{ss} = \Omega Q^{\frac{1}{2}}/(gH_s)^{\frac{3}{4}}$, where $\Omega = U_{s1}/r_{s1}$ and $Q = c_{x1}A_1 = \pi k r_{s1}^2 c_{x1}$. Thus,

$$\frac{\Omega_{ss}^2}{\pi k} = \frac{U_{s1}^2 c_{x1}}{\{\frac{1}{2}c_{x1}^2(1 + \sigma_b) + \frac{1}{2}\sigma_b U_{s1}^2\}^{3/2}} = \frac{\phi}{\{\frac{1}{2}(1 + \sigma_b)\phi^2 + \frac{1}{2}\sigma_b\}^{3/2}}$$

(7.7)

where $\phi = c_{x1}/U_{s1}$. To obtain the condition of maximum Ω_{ss}, eqn. (7.7) is differentiated with respect to ϕ and the result set equal to zero. From this procedure the optimum conditions are found:

$$\phi = \left\{\frac{\sigma_b}{2(1 + \sigma_b)}\right\}^{\frac{1}{2}},$$

(7.8a)

$$gH_s = \frac{3}{2}\sigma_b(\frac{1}{2}U_{s1}^2),$$

(7.8b)

$$\Omega_{ss}^2 = \frac{2\pi k(2/3)^{1\cdot5}}{\sigma_b(1 + \sigma_b)^{0\cdot5}} = \frac{3\cdot420k}{\sigma_b(1 + \sigma_b)^{0\cdot5}}.$$

(7.8c)

EXAMPLE: The inlet of a centrifugal pump of the type shown in Fig. 7.1 is to be designed for optimum conditions when the flow rate of water is 25 dm³/s and the impeller rotational speed is 1450 rev/min. The maximum suction specific speed $\Omega_{ss} = 3\cdot0$ (rad) and the inlet eye radius ratio is to be 0·3. Determine

   (i) the blade cavitation coefficient,
  (ii) the shroud diameter at the eye,
 (iii) the eye axial velocity, and
 (iv) the NPSH.

Solution: (i) From eqn. (7.8c),

$$\sigma_b^2(1 + \sigma_b) = (3\cdot42\ k)^2/\Omega_{ss}^4 = 0\cdot1196$$

with $k = 1 - (r_{h1}/r_{s1})^2 = 1 - 0\cdot3^2 = 0\cdot91$. Solving iteratively (e.g. using the Newton–Raphson approximation), $\sigma_b = 0\cdot3030$.

(ii) As $Q = \pi k r_{s1}^2 c_{x1}$ and $c_{x1} = \phi\Omega r_{s1}$

then $r_{s1}^3 = Q/(\pi k\Omega\phi)$ and $\Omega = 1450\pi/30 = 151\cdot84$ rad/s.

From eqn. (7.8a), $\phi = \{0\cdot303/(2 \times 1\cdot303)\}^{0\cdot5} = 0\cdot3410$,

$$\therefore r_{s1}^3 = 0\cdot025/(\pi \times 0\cdot91 \times 151\cdot84 \times 0\cdot341) = 1\cdot689 \times 10^{-4},$$

$$\therefore r_{s1} = 0\cdot05528 \text{ m.}$$

The required diameter of the eye is 110·6 mm.

(iii) $c_{x1} = \phi\Omega r_{s1} = 0\cdot341 \times 151\cdot84 \times 0\cdot05528 = 2\cdot862$ m/s.

(iv) From eqn. (7.8b),

$$H_s = \frac{0\cdot75\sigma_b c_{x1}^2}{g\phi^2} = \frac{0\cdot75 \times 0\cdot303 \times 2\cdot862^2}{9\cdot81 \times 0\cdot341^2} = 1\cdot632 \text{ m.}$$

OPTIMUM DESIGN OF A CENTRIFUGAL COMPRESSOR INLET

To obtain high efficiencies from high pressure ratio compressors it is necessary to limit the relative Mach number at the eye.

The flow area at the eye can be written as

$$A_1 = \pi r_{s1}^2 k, \quad \text{where } k = 1 - (r_{h1}/r_{s1})^2.$$

Hence $A_1 = \pi k U_{s1}^2/\Omega^2$ (7.9)

with $U_{s1} = \Omega r_{s1}.$

With uniform axial velocity the continuity equation is $\dot{m} = \rho_1 A_1 c_{x1}$.
Noting from the inlet velocity diagram (Fig. 7.1) that $c_{x1} = w_{s1} \cos \beta_{s1}$ and $U_{s1} = w_{s1} \sin \beta_{s1}$, then, using eqn. (7.9),

$$\frac{\dot{m}\Omega^2}{\rho_1 k\pi} = w_{s1}^3 \sin^2 \beta_{s1} \cos \beta_{s1}. \quad\quad (7.10)$$

For a perfect gas it is most convenient to express the static density ρ_1 in terms of the stagnation temperature T_{01} and stagnation pressure p_{01} because these parameters are usually constant at entry to the compressor. Now,

$$\frac{\rho}{\rho_0} = \frac{p}{p_0} \frac{T_0}{T}.$$

With $C_p T_0 = C_p T + \frac{1}{2}c^2$ and $C_p = \gamma R/(\gamma - 1)$

then

$$\frac{T_0}{T} = 1 + \frac{\gamma - 1}{2} M^2 = \frac{a_0{}^2}{a^2}$$

where the Mach number, $M = c/(\gamma RT)^{\frac{1}{2}} = c/a$, a_0 and a being the stagnation and local (static) speeds of sound. For isentropic flow,

$$\frac{p}{p_0} = \left(\frac{T}{T_0}\right)^{\gamma/(\gamma-1)}.$$

Thus,

$$\frac{\rho_1}{\rho_0} = \left(\frac{T_1}{T_0}\right)^{1-\gamma/(\gamma-1)} = \left(1 + \frac{\gamma - 1}{2} M_1{}^2\right)^{-1/(\gamma-1)}$$

where

$$\rho_0 = p_0/(RT_0).$$

The absolute Mach number M_1 and the relative Mach number M_{r1} are defined as

$$M_1 = c_{x1}/a_1 = M_{r1} \cos \beta_{s1} \quad \text{and} \quad w_{s1} = M_{r1}a_1.$$

Using these relations together with eqn. (7.10)

$$\frac{\dot{m}\Omega^2 RT_{01}}{k\pi p_{01}} = \frac{M_{r1}^3 a_1^3}{[1 + \frac{1}{2}(\gamma - 1)M_1^2]^{1/(\gamma-1)}} \sin^2 \beta_{s1} \cos \beta_{s1}$$

Since $a_{01}/a_1 = [1 + \frac{1}{2}(\gamma-1)M_1^2]^{\frac{1}{2}}$ and $a_{01} = (\gamma RT_{01})^{\frac{1}{2}}$ the above equation is rearranged to give

$$\frac{\dot{m}\Omega^2}{\pi k\gamma p_{01}(\gamma RT_{01})^{\frac{1}{2}}} = \frac{M_{r1}^3 \sin^2 \beta_{s1} \cos \beta_{s1}}{[1 + \frac{1}{2}(\gamma - 1)M_{r1}^2 \cos^2 \beta_{s1}]^{1/(\gamma-1)+\frac{3}{2}}} \qquad (7.11)$$

This equation is extremely useful and can be used in a number of different ways. For a particular gas and known inlet conditions one can specify values of γ, R, p_{01} and T_{01} and obtain $\dot{m}\Omega^2/k$ as a function of M_{r1} and β_{s1}. By specifying a particular value of M_{r1} as a limit, the optimum value of β_{s1} for maximum mass flow can be found. A graphical procedure is the simplest method of optimising β_{s1} as illustrated below.

Taking as an example air ($\gamma = 1\cdot4$, $R = 287$ J/(kg°C)) entering the compressor at the stagnation conditions $p_{01} = 101\cdot3$ kPa and $T_{01} = 288$ K, then eqn. (7.11) becomes

$$0.6598 \times 10^{-8}\frac{\dot{m}\Omega^2}{k} = \frac{M_{r1}^3 \sin^2 \beta_{s1} \cos \beta_{s1}}{[1 + \frac{1}{5}M_{r1}^2 \cos^2 \beta_{s1}]^4} \qquad (7.12)$$

The right-hand side of eqn. (7.12) is plotted in Fig. 7.4 as a function of

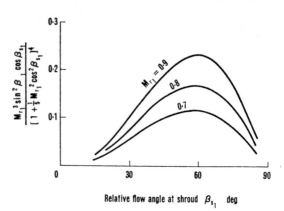

Relative flow angle at shroud β_{s_1} deg

FIG. 7.4. Mass flow function for a centrifugal compressor with zero entry swirl (adapted from Shepherd[5]).

β_{s1} for several discrete values of M_{r1}. These curves are all seen to peak at approximately $\beta_{s1} = 60$ deg at which condition $\dot{m}\Omega^2/k$ is a maximum.

Shepherd[5] has analysed this problem with a more general approach which includes the effect of inlet swirl ($c_{\theta 1} \neq 0$). His results show that a *prerotation* of the fluid which decreases the relative velocity, increases the peak values of $\dot{m}\Omega^2/k$ significantly and reduces the values of β_{s1} at which the peaks occur.

PREWHIRL

Prerotation or prewhirl can be induced in the entry flow of a pump or compressor by fitting guide vanes in the inlet section. One method of doing this is shown in Fig. 7.5a. The object is to reduce the relative velocity to the impeller and Fig. 7.5b shows the effect vectorially.

For guide vanes designed to produce a free-vortex flow, the axial velocity is constant (in the absence of viscous effects) and the tangential velocity component c_θ varies inversely with radius. Thus a

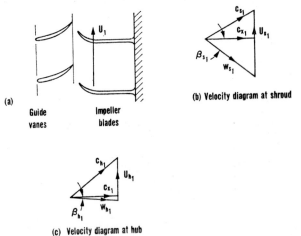

(a)

Guide   Impeller
vanes   blades

(b) Velocity diagram at shroud

(c) Velocity diagram at hub

FIG. 7.5. Effect of free-vortex prewhirl vanes upon relative velocity at impeller inlet.

prewhirl at the eye tip sufficient to reduce the relative velocity w_{s1} to an acceptable value also has the effect of increasing the swirl at the hub which, together with the lower blade speed, results in a very low or even zero relative flow angle β_{h1}. This effect is made clear by a comparison of the velocity diagrams of Fig. 7.5b and c. This may have certain advantages in some methods of impeller manufacture.

One obvious disadvantage in the use of prewhirl is that the energy transfer is reduced by the amount $U_1 c_{\theta 1}$. It will be noted that this is a constant quantity for a free-vortex distribution since $U_1 \propto r$ and $c_{\theta 1} \propto 1/r$.

SLIP FACTOR

Even under ideal frictionless conditions the relative flow leaving a compressor or pump impeller receives less than perfect guidance from the vanes and the flow is said to *slip*. Figure 7.6 compares the relative flow angle β_2 with the impeller tip *vane* angle β_2'. A *slip factor* σ may be defined as $c_{\theta 2}/c_{\theta 2}'$, where $c_{\theta 2}$, $c_{\theta 2}'$ are the tangential components of absolute velocity corresponding to the angles β_2, β_2' respectively. The slip factor is a piece of information of vital importance to the compressor

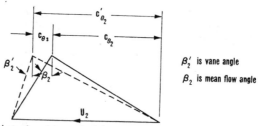

FIG. 7.6. Actual and hypothetical velocity diagrams at outlet from a "backward leaning" impeller.

designer as accurate knowledge of it enables the energy transfer between the impeller and fluid to be found.

Many attempts at predicting the amount of slip from impellers have been made and a useful summary of some of this work is given by Stanitz.[6] Most of the quantitative results are for radial blades ($\beta_2' = 0$ deg) although results for backward leaning blades ($\beta_2' > 0$ deg) and for some mixed flow impellers have also been given. All these analyses are based on the assumption of an inviscid fluid.

If a frictionless fluid passes through an impeller without spin, then at outlet the spin *must still be zero*. The impeller itself has an angular velocity Ω so that, relative to the impeller, the fluid must have an angular velocity of $-\Omega$; this quantity is termed the *relative eddy*. A most simple explanation for the slip effect in an impeller is obtained from the idea of relative eddy.

At the impeller outlet the relative flow can be regarded as a through flow on which is superimposed a relative eddy. The net effect of these two motions is that the flow at outlet is inclined to the vane tip in the direction away from the blade motion (Fig. 7.7).

One of the earliest and simplest expressions for slip was obtained by Stodola.[7] Referring to Fig. 7.8 the *slip velocity*, $c_{\theta s} = c_{\theta 2}' - c_{\theta 2}$, is considered to be the product of the relative eddy and the radius $d/2$ of a circle which can be inscribed within the channel. Thus $c_{\theta s} = \Omega d/2$. If the number of vanes is denoted by Z then an approximate expression, $d \simeq (2\pi r_2/Z) \cos \beta_2'$ can be written if Z is not small. Since $\Omega = U_2/r_2$ then

$$c_{\theta s} = \frac{\pi U_2 \cos \beta_2'}{Z}. \qquad (7.13)$$

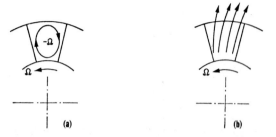

Fig. 7.7. (a) Relative eddy without any throughflow. (b) Relative flow at impeller exit (throughflow added to relative eddy).

Now as $c'_{\theta 2} = U_2 - c_{r2} \tan \beta'_2$ the Stodola slip factor becomes

$$\sigma = \frac{c_{\theta 2}}{c'_{\theta 2}} = 1 - \frac{c_{\theta s}}{U_2 - c_{r2} \tan \beta'_2} \tag{7.14}$$

or,

$$\sigma = 1 - \frac{(\pi/Z) \cos \beta'_2}{1 - \phi_2 \tan \beta'_2} \tag{7.15}$$

where $\phi_2 = c_{r2}/U_2$.

A number of more refined (mathematically exact) solutions have been evolved of which the most well known are those of Busemann, discussed at some length by Wislicenus[8] and Stanitz[6] mentioned earlier. The volume of mathematical work required to describe these theories is too extensive to justify inclusion here and only a brief outline of the results is presented.

Busemann's theory applies to the special case of two-dimensional vanes curved as logarithmic spirals as shown in Fig. 7.9. Considering the geometry of the vane element shown it should be an easy task for the student to prove that,

$$\gamma = \tan \beta' \ln (r_2/r_1) \tag{7.17a}$$

that the ratio of vane length to equivalent blade pitch is

$$\frac{l}{s} = \frac{Z}{2 \pi \cos \beta'} \ln \left(\frac{r_2}{r_1}\right) \tag{7.17b}$$

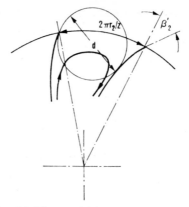

Fɪɢ. 7.8. Flow model for Stodola slip factor.

and that the equivalent pitch is

$$s = \frac{2\pi(r_2 - r_1)}{Z \ln (r_2/r_1)}.$$

The equi-angular or logarithmic spiral is the simplest form of radial vane system and has been frequently used for *pump impellers* in the past. The Busemann slip factor can be written as

$$\sigma_B = (A - B\phi_2 \tan \beta_2')/(1 - \phi_2 \tan \beta_2'), \qquad (7.16)$$

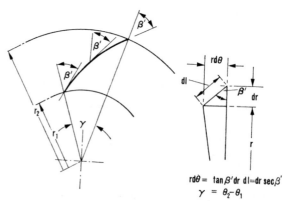

Fɪɢ. 7.9. Logarithmic spiral vane. Vane angle β' is constant for all radii.

where both A and B are functions of r_2/r_1, β_2' and Z. For typical pump and compressor impellers the dependence of A and B on r_2/r_1 is negligible when the equivalent l/s exceeds unity. From eqn. (7.17b) the requirement for $l/s \geq 1$, is that the radius ratio must be sufficiently large, i.e.

$$r_2/r_1 \geq \exp(2\pi \cos \beta'/Z). \tag{7.17c}$$

This criterion is often applied to other than logarithmic spiral vanes and then β_2' is used instead of β'. Radius ratios of typical centrifugal pump impeller vanes normally exceed the above limit. For instance, blade outlet angles of impellers are usually in the range $50 \leqslant \beta_2' \leqslant 70$ deg with between 5 and 12 vanes. Taking representative values of $\beta_2' = 60$ deg and $Z = 8$ the R.H.S. of eqn. (7.17c) is equal to 1·48 which is not particularly large for a pump.

So long as these criteria are obeyed the value of B is constant and practically equal to unity for all conditions. Similarly, the value of A

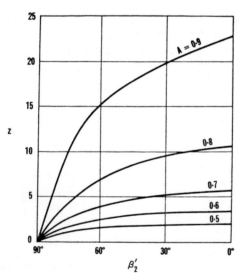

FIG. 7.10. Head correction factors for centrifugal impellers (adapted from Csanady[10]).

is independent of the radius ratio r_2/r_1 and depends on β_2' and Z only. Values of A are shown in Fig. 7.10[10] and may also be interpreted as the value of σ_B for zero through flow ($\phi_2 = 0$).

The exact solution of Busemann makes it possible to check the validity of approximate methods of calculation such as the Stodola expression. By putting $\phi_2 = 0$ in eqns. (7.15) and (7.16) a comparison of the Stodola and Busemann slip factors at the zero through flow condition can be made. The Stodola value of slip comes close to the exact correction if the vane angle is within the range $50 \leqslant \beta_2' \leqslant 70$ deg and the number of vanes exceeds 6.

Stanitz[6] applied relaxation methods of calculation to solve the potential flow field between the blades (blade-to-blade solution) of eight impellers with blade tip angles β_2' varying between 0 and 45 deg. His main conclusions were that the computed slip velocity $c_{\theta s}$ was independent of vane angle β_2' and depended only on blade spacing (number of blades). He also found that compressibility effects did not affect the slip factor. Stanitz's expression for slip velocity is,

$$c_{\theta s} = 0 \cdot 63 U_2 \pi / Z \qquad (7.18)$$

and the corresponding slip factor σ_s using eqn. (7.14) is

$$\sigma_s = 1 - \frac{0 \cdot 63\pi/Z}{1 - \phi_2 \tan \beta_2'}. \qquad (7.18a)$$

For radial vaned impellers this becomes $\sigma_s = 1 - 0 \cdot 63\pi/Z$ but is often written for convenience and initial approximate calculations as $\sigma_s = 1 - 2/Z$.

Ferguson[9] has usefully compiled values of slip factor found from several theories for a number of blade angles and blade numbers and compared them with known experimental values. He found that for pumps, with β_2' between 60 deg and 70 deg, the Busemann or Stodola slip factors gave fairly good agreement with experimental results. For radial vaned impellers on the other hand, the Stanitz expression, eqn. (7.18a) agreed very well with experimental observations. For intermediate values of β_2' the Busemann slip factor gave the most consistent agreement with experiment.

HEAD INCREASE OF A CENTRIFUGAL PUMP

The actual delivered head H measured as the *head difference* between the inlet and outlet flanges of the pump and sometimes called the *manometric head*, is less than the ideal head H_i defined by eqn. (7.1b) by the amount of the internal losses. The hydraulic efficiency of the pump is defined as

$$\eta_h = \frac{H}{H_i} = \frac{gH}{U_2 c_{\theta 2}}. \tag{7.19}$$

From the velocity triangles of Fig. 7.2

$$c_{\theta 2} = U_2 - c_{r2} \tan \beta_2.$$

Therefore $\quad H = \eta_h \, U_2{}^2 \, (1 - \phi_2 \tan \beta_2)/g \tag{7.19a}$

where $\phi_2 = c_{r2}/U_2$ and β_2 is the actual averaged relative flow angle at impeller outlet.

With the definition of slip factor, $\sigma = c_{\theta 2}/c_{\theta 2'}$, H can, more usefully, be directly related to the impeller vane outlet angle, as

$$H = \eta_h \sigma U_2{}^2 (1 - \phi_2 \tan \beta_2')/g. \tag{7.19b}$$

In general, centrifugal pump impellers have between five and twelve vanes inclined backwards to the direction of rotation, as suggested in Fig. 7.2, with a vane tip angle β_2' of between 50 and 70 deg. A knowledge of blade number, β_2' and ϕ_2 (usually small and of the order 0·1) generally enables σ to be found using the Busemann formula. The effect of slip, it should be noted, causes the relative flow angle β_2 to become larger than the vane tip angle β_2'.

EXAMPLE: A centrifugal pump delivers 0·1 m^3/s of water at a rotational speed of 1200 rev/min. The impeller has seven vanes which lean backwards to the direction of rotation such that the vane tip angle β_2' is 50 deg. The impeller has an external diameter of 0·4 m, an internal diameter of 0·2 m and an axial width of 31·7 mm. Assuming that the diffuser efficiency is 51·5%, that the impeller head losses are 10% of the ideal head rise and that the diffuser exit is 0·15 m in diameter, estimate the slip factor, the manometric head and the hydraulic efficiency.

Solution. Equation (7.16) is used for estimating the slip factor. Since $\exp(2\pi \cos \beta_2'/Z) = \exp (2\pi \times 0·643/7) = 1·78$, is less than $r_2/r_1 = 2$, then $B = 1$ and $A \simeq 0.77$, obtained by replotting the values of A given in Fig. 7.10 for $\beta_2' = 50$ deg and interpolating.

The vane tip speed, $U_2 = \pi N D_2/60 = \pi \times 1200 \times 0.4/60 = 25.13$ m/s.
The radial velocity, $c_{r2} = Q/(\pi D_2 b_2) = 0.1/(\pi \times 0.4 \times 0.0317)$
$= 2.51$ m/s.

Hence the Busemann slip factor is

$$\sigma_B = (0.77 - 0.1 \times 1.192)/(1 - 0.1 \times 1.192) = 0.739.$$

Hydraulic losses occur in the impeller and in the diffuser. The kinetic energy leaving the diffuser is not normally recovered and must contribute to the total loss, H_L. From inspection of eqn. (2.48), the loss in head in the diffuser is $(1-\eta_D)(c_2{}^2-c_3{}^2)/(2g)$. The head loss in the impeller is $0.1 \times U_2 c_{\theta2}/g$ and the exit head loss is $c_3{}^2/(2g)$. Summing the losses,

$$H_L = 0.485\ (c_2{}^2-c_3{}^2)/(2g) + 0.1 \times U_2 c_{\theta2}/g + c_3{}^2/(2g).$$

Determining the velocities and heads needed,

$c_{\theta2} = \sigma_B U_2 (1 - \phi_2 \tan \beta_2') = 0.739 \times 25.13 \times 0.881 = 16.35$ m/s.
$H_i = U_2 c_{\theta2}/g = 25.13 \times 16.35/9.81 = 41.8$ m.
$c_2{}^2/(2g) = (16.35^2 + 2.51^2)/19.62 = 13.96$ m.
$c_3 = 4Q/(\pi d^2) = 0.4/(\pi \times 0.15^2) = 5.65$ m/s.
Therefore $c_3{}^2/(2g) = 1.63$ m.
Therefore $H_L = 4.18 + 0.485\ (13.96 - 1.63) + 1.63 = 11.8$ m.

The manometric head is

$$H = H_i - H_L = 41.8 - 11.8 = 30.0\ \text{m}$$

and the hydraulic efficiency

$$\eta_h = H/H_i = 71.7\%.$$

PRESSURE RATIO OF A CENTRIFUGAL COMPRESSOR

Consider a centrifugal compressor having zero inlet swirl, compressing a perfect gas. With the usual notation the energy transfer is

$$\Delta W = \dot{W}_c/\dot{m} = h_{02} - h_{01} = U_2 c_{\theta2}.$$

The overall or total to total efficiency η_c is

$$\eta_c = \frac{h_{03ss} - h_{01}}{h_{03} - h_{01}} = \frac{C_p T_{01}(T_{03ss}/T_{01} - 1)}{h_{02} - h_{01}}$$

$$= C_p T_{01}(T_{03ss}/T_{01} - 1)/(U_2 c_{\theta 2}). \qquad (7.20)$$

Now the overall pressure ratio is

$$\frac{p_{03}}{p_{01}} = \left(\frac{T_{03ss}}{T_{01}}\right)^{\gamma/(\gamma-1)}. \qquad (7.21)$$

Substituting eqn. (7.20) into eqn. (7.21) and noting that $C_p T_{01} = \gamma R T_{01}/(\gamma - 1) = a_{01}^2/(\gamma - 1)$, the pressure ratio becomes

$$\frac{p_{03}}{p_{01}} = \left[1 + \frac{(\gamma - 1)\eta_c U_2 c_{r2} \tan \alpha_2}{a_{01}{}^2}\right]^{\gamma/(\gamma-1)}. \qquad (7.22)$$

From the velocity triangle at impeller outlet (Fig. 7.1)

$$\phi_2 = c_{r2}/U_2 = (\tan \alpha_2 + \tan \beta_2)^{-1}$$

and, therefore,

$$\frac{p_{03}}{p_{01}} = \left[1 + \frac{(\gamma - 1)\eta_c U_2^2 \tan \alpha_2}{a_{01}^2(\tan \alpha_2 + \tan \beta_2)}\right]^{\gamma/(\gamma-1)}. \qquad (7.23a)$$

This formulation is useful if the flow angles can be specified. Alternatively, as $c_{r2} \tan \alpha_2 = U_2(1 - \phi_2 \tan \beta_2)$,

$$\frac{p_{03}}{p_{01}} = [1 + (\gamma - 1)\eta_c U_2^2(1 - \phi_2 \tan \beta_2)/a_{01}^2]^{\gamma/(\gamma-1)}. \qquad (7.23b)$$

The blade tip speed U_2 of high-pressure ratio compressors must, of necessity, be high and non-radial blades would be subjected to large bending stresses as a result of centrifugal forces. For such high-performance compressors, radial blades are advocated ($\beta_2' = 0$) and the following simple relation for pressure ratio is obtained:

$$\frac{p_{03}}{p_{01}} = [1 + (\gamma - 1)\eta_c U_2^2 \sigma/a_{01}^2]^{\gamma/(\gamma-1)}, \qquad (7.23c)$$

it being seen that $c_{\theta 2} = \sigma U_2$, where the Stanitz relation for the slip factor applies, $\sigma_s = 1 - 2/Z$.

It is of some interest to calculate the pressure ratio of a radial-bladed centrifugal air compressor, using eqn. (7.23c), with typical values of blade tip speed and efficiency. Figure 7.11 shows the variation

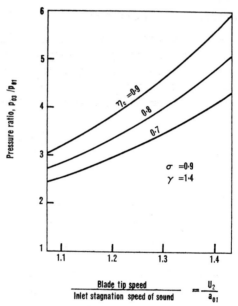

Fig. 7.11. Variation of pressure ratio with blade speed for a radial-bladed compressor ($\beta_2' = 0$) at various values of efficiency.

of pressure ratio with blade tip speed for three values of compressor efficiency with air taken at normal conditions ($a_{01} = 340{\cdot}5$ m/s and $\gamma = 1{\cdot}4$). The slip factor is taken as 0·9 representing a typical compressor having nineteen radial blades. It is clear from Fig. 7.11 the strong effect of both efficiency and blade speed on pressure ratio. The limit on blade speed due to centrifugal stresses appears to be about 500 m/s and efficiencies of centrifugal compressors seldom exceed 80% so that the highest pressure ratios obtained in one stage are about 5 to 1.

It appears characteristic of all compressors that the range of mass flow between surge and choke becomes smaller the higher the pressure ratio for which the machine is designed. In the case of a centrifugal compressor, choking normally occurs when the Mach number entering the diffuser passages is about unity. The problem is complicated by shock-induced separation of boundary layer on the vanes which tends to aggravate the flow blockage problem.

MACH NUMBER AT IMPELLER EXIT

With high blade tip speeds the absolute flow leaving the impeller may have a Mach number well in excess of unity. As this Mach number can be related to the Mach number at entry to the diffuser vanes, it is of some advantage to be able to calculate the former.

Assuming a perfect gas the Mach number at impeller exit M_2 can be written as

$$M_2^2 = \frac{c_2^2}{a_2^2} = \frac{c_2^2}{T_{01}} \cdot \frac{T_{01}}{T_2} \cdot \frac{T_2}{a_2^2} = \frac{c_2^2}{a_{01}^2} \frac{T_{01}}{T_2}, \qquad (7.24)$$

since $a_{01}^2 = \gamma R T_{01}$ and $a_2^2 = \gamma R T_2$.

Referring to the outlet velocity triangle, Fig. 7.1,

$$c_2^2 = c_{r2}^2 + (U_2 - c_{r2} \tan \beta_2)^2$$
$$= U_2^2[\phi_2^2 + (1 - \phi_2 \tan \beta_2)^2]. \qquad (7.25)$$

Assuming zero inlet swirl the inlet velocity triangle can be used to obtain $h_{01} = h_1 + \frac{1}{2}(w_1^2 - U_1^2)$. Inserting this in eqn. (7.4)

$$h_2 = h_{01} + \frac{1}{2}(U_2^2 - w_2^2)$$

hence

$$T_2/T_{01} = 1 + (1 - \phi_2^2 \sec^2 \beta_2)U_2^2/(2C_p T_{01}). \qquad (7.26)$$

Substituting eqns. (7.25), (7.26) into eqn. (7.24)

$$M_2^2 = \frac{U_2^2}{a_{01}^2} \frac{[\phi_2^2 + (1 - \phi_2 \tan \beta_2)^2]}{[1 + \frac{1}{2}(\gamma - 1)(U_2^2/a_{01}^2)(1 - \phi_2^2 \sec^2 \beta_2)]}, \qquad (7.27)$$

noting that $C_p T_{01} = a_{01}^2/(\gamma - 1)$.

For the important case of radial vaned impellers $(\beta_2' = 0)$, $\sigma = 1 - \phi_2 \tan \beta_2$, which can be used in eqn. (7.27). After some rearranging,

$$M_2^2 = \frac{U_2^2(\phi_2^2 + \sigma^2)}{a_{01}^2[1 + \frac{1}{2}(\gamma - 1)(U_2^2/a_{01}^2)\{\sigma(2 - \sigma) - \phi_2^2\}]}. \qquad (7.28)$$

With $U_2 = 500$ m/s, $\sigma = 0.9$, $\gamma = 1.4$ and $a_{01} = 340.5$ m/s, eqn. (7.28) can be used to find M_2 if ϕ_2 is specified. Since blade tip speeds are high, ϕ_2 is usually low and the Mach number M_2 is relatively insensitive to ϕ_2. For $0.1 \leqq \phi_2 \leqq 0.3$, M_2 varies in the range $1.2 \geqq M_2 \geqq 1.16$.

EXAMPLE: Air at a stagnation temperature of 22°C enters the impeller of a centrifugal compressor in the axial direction. The rotor, which has 17 radial vanes, rotates at 15,000 rev/min. The stagnation pressure ratio between diffuser outlet and impeller inlet is 4·2 and the overall efficiency (total to total) is 83%. Determine the impeller tip radius and power required to drive the compressor when the mass flow rate is 2 kg/s and the mechanical efficiency is 97%. Given that the air density at impeller outlet is 2 kg/m³ and the axial width at entrance to the diffuser is 11 mm, determine the absolute Mach number at that point. Assume that the slip factor $\sigma_s = 1 - 2/Z$, where Z is the number of vanes.

(For air take $\gamma = 1.4$ and $R = 287$ J/kg°C.)

Solution. From eqn. (7.1a) the specific work is

$$\Delta W = h_{02} - h_{01} = U_2 c_{\theta 2} = \sigma_s U_2{}^2$$

since $c_{\theta 1} = 0$. Combining eqns. (7.20) and (7.21) with the above and rearranging gives

$$U_2{}^2 = \frac{C_p T_{01}(r^{(\gamma-1)/\gamma} - 1)}{\sigma_s \eta_c}$$

where $r = p_{03}/p_{01} = 4.2$; $C_p = \gamma R/(\gamma - 1) = 1005$ J/kg°C; $\sigma_s = 1 - 2/17$
$$= 0.8824.$$

Therefore $U_2{}^2 = \dfrac{1005] \times 295(4.2^{0.286} - 1)}{0.8824 \times 0.83} = 20.5 \times 10^4.$

Therefore $U_2 = 452$ m/s.

The rotational speed is

$$\Omega = 15,000 \times 2\pi/60 = 1570 \text{ rad/s.}$$

Thus, the impeller tip radius is

$$r_t = U_2/\Omega = 452/1570 = 0.288 \text{ m.}$$

The actual shaft power is obtained from

$$\dot{W}_{act} = \dot{W}_c/\eta_m = \dot{m}\Delta W/\eta_m = 2 \times 0{\cdot}8824 \times 452^2/0{\cdot}97$$
$$= 373 \text{ kW.}$$

Although the absolute Mach number at the impeller tip can be obtained almost directly from eqn. (7.28) it may be instructive to find it from

$$M_2 = \frac{c_2}{a_2} = \frac{c_2}{(\gamma R T_2)^{\frac{1}{2}}}$$

where $c_2 = (c_{\theta 2}{}^2 + c_{r2}{}^2)^{\frac{1}{2}}.$

$$c_{r2} = \dot{m}/(\rho_2 2\pi r_t b_2) = 2/(2 \times 2\pi \times 0{\cdot}288 \times 0{\cdot}011) = 50{\cdot}3 \text{ m/s}$$

$$c_{\theta 2} = \sigma_s U_2 = 400 \text{ m/s.}$$

Therefore $c_2 = \sqrt{(400^2 + 50{\cdot}3^2)} = 402{\cdot}5$ m/s.

Since $h_{02} = h_{01} + \Delta W$

$$h_2 = h_{01} + \Delta W - \tfrac{1}{2}c_2{}^2.$$

Therefore $T_2 = T_{01} + (\Delta W - \tfrac{1}{2}c_2{}^2)/C_p = 295 + (18{\cdot}1 - 8{\cdot}1)10^4/1005$
$$= 394{\cdot}5 \text{ K.}$$

Hence,

$$M_2 = \frac{402{\cdot}5}{\sqrt{(402 \times 394{\cdot}5)}} = 1{\cdot}01.$$

THE DIFFUSER SYSTEM

Centrifugal compressors and pumps are, in general, fitted with either a vaneless or a vaned diffuser to transform the kinetic energy at impeller outlet into static pressure.

Vaneless diffusers

The simplest concept of diffusion in a radial flow machine is one where the swirl velocity is reduced by an increase in radius (conservation of angular momentum) and the radial velocity component is controlled by the radial flow area. From continuity, since $\dot{m} = \rho A c_r = 2\pi b \rho c_r$, where b is the width of passage, then

$$c_t = \frac{r_2 b_2 \rho_2 c_{r2}}{rb\rho} . \qquad (7.29)$$

Assuming the flow is frictionless in the diffuser, the angular momentum is constant and $c_\theta = c_{\theta 2} r_2/r$. Now tangential velocity component c_θ is usually very much larger than the radial velocity component c_r; therefore, the ratio of inlet to outlet diffuser velocities c_2/c_3 is approximately r_3/r_2. Clearly, to obtain useful reductions in velocity, vaneless diffusers must be large. This may not be a disadvantage in industrial applications where weight and size may be of secondary importance compared with the cost of a vaned diffuser. A factor in favour of vaneless diffusers is the wide operating range obtainable, vaned diffusers being more sensitive to flow variation because of incidence effects.

For a parallel walled radial diffuser in incompressible flow, rc_r is constant and, therefore, $\tan \alpha = c_\theta/c_r = $ constant. Under these conditions the flow maintains a constant inclination α to radial lines and the flow path traces a *logarithmic spiral*. The law relating radius ratio to the change in angle θ (Fig. 7.9) can be found from the flow geometry as follows. For an increment in radius, dr, $rd\theta = dr \tan \alpha$ which is integrated to give

$$\theta - \theta_2 = \tan \alpha \, \log(r_2/r_1). \qquad (7.30)$$

Taking $\alpha = 78$ deg and $r_3/r_2 = 2$ to be fairly representative values, the change in angle, $\theta_3 - \theta_2$ is nearly 180 deg.

Because of the long flow path with this type of diffuser, friction effects are important and the efficiency is low.

Vaned diffusers

In the vaned diffuser the vanes are used to remove the swirl of the fluid at a higher rate than is possible by a simple increase in radius, thereby reducing the length of flow path and diameter. The vaned diffuser is advantageous where small size is important.

A typical vaned diffuser configuration is illustrated in Fig. 7.1. There is a clearance between the impeller and vane leading edges amounting to about $0.04 \, D_2$ for pumps and between $0.1 \, D_2$ to $0.2 \, D_2$

for compressors. This space constitutes a vaneless diffuser and its functions are (i) to reduce the circumferential pressure gradient at the impeller tip, (ii) to smooth out velocity variations between the impeller tip and vanes, and (iii) to reduce the Mach number (for compressors) at entry to the vanes.

The flow follows an approximately logarithmic spiral path to the vanes after which it is constrained by the diffuser channels. For rapid diffusion the axis of the channel is straight and tangential to the spiral as shown. The passages are generally designed on the basis of simple channel theory with an equivalent angle of divergence of between 8 deg and 12 deg to control separation. (See remarks in Chapter 2 on straight-walled diffuser efficiency.)

The number of diffuser vanes has a direct bearing on the size and efficiency of the vaneless diffuser. With a large number of vanes, the angle of divergence is smaller and the diffuser becomes more efficient up to the point where increased friction and blockage overcomes the advantage of more gradual diffusion. Also, regarding the vanes as a blade cascade, an increased number of vanes implies a reduced radius ratio since the space–chord ratio will be roughly constant. However, as discussed by Cheshire,[1] too many diffuser passages can have a strong adverse effect on the surge characteristics of a centrifugal compressor.

With several adjacent diffuser passages sharing the gas from one impeller passage, the uneven velocity distribution from that passage results in alternate diffuser passages being either starved or choked. This is an unstable situation leading to flow reversal in the passages and to surge of the compressor. When the number of diffuser passages is *less* than the number of impeller passages a more uniform total flow results.

CHOKING IN A COMPRESSOR STAGE

When the through flow velocity in a passage reaches the speed of sound at some cross-section, the flow *chokes*. For the stationary inlet passage this means that no further increase in mass flow is possible, either by decreasing the back pressure or by increasing the rotational speed. Now the choking behaviour of rotating passages differs from that of stationary passages, making separate analyses for the inlet, impeller and diffuser a necessity. For each component a simple, one-dimensional approach is used assuming that all flow processes are adiabatic and that the fluid is a perfect gas.

Inlet

Choking takes place when $c^2 = a^2 = \gamma RT$. Since $h_0 = h + \frac{1}{2}c^2$, then $C_p T_0 = C_p T + \frac{1}{2}\gamma RT$ and

$$\frac{T}{T_0} = \left(1 + \frac{\gamma R}{2C_p}\right)^{-1} = \frac{2}{\gamma + 1}. \qquad (7.31)$$

Assuming the flow in the inlet is isentropic,

$$\frac{\rho}{\rho_0} = \frac{p}{p_0}\frac{T_0}{T} = [1 + \frac{1}{2}(\gamma - 1)M^2]^{1 - \gamma/(\gamma - 1)}$$

and when $c = a$, $M = 1$, so that

$$\frac{\rho}{\rho_0} = \left(\frac{2}{\gamma + 1}\right)^{1/(\gamma - 1)} \qquad (7.32)$$

Substituting eqns. (7.31), (7.32) into the continuity equation, $\dot{m}/A = \rho c = \rho(\gamma RT)^{\frac{1}{2}}$, then

$$\frac{\dot{m}}{A} = \rho_0 a_0 \left(\frac{2}{\gamma + 1}\right)^{(\gamma + 1)/2(\gamma - 1)}. \qquad (7.33)$$

Thus, since ρ_0, a_0 refer to inlet stagnation conditions which remain unchanged, the mass flow rate at choking is constant.

Impeller

In the rotating impeller passages, flow conditions are referred to the factor $I = h + \frac{1}{2}(w^2 - U^2)$, which is constant according to eqn. (7.4). At the impeller inlet and for the special case $c_{\theta 1} = 0$, note that $I_1 = h_1 + \frac{1}{2}c_1^2 = h_{01}$. When choking occurs in the impeller passages it is the *relative velocity* w which equals the speed of sound at some section. Now $w^2 = a^2 = \gamma RT$ and $T_{01} = T + (\gamma RT/2C_p) - (U^2/2C_p)$, therefore

$$\frac{T}{T_{01}} = \left(\frac{2}{\gamma + 1}\right)\left(1 + \frac{U^2}{2C_pT_{01}}\right). \tag{7.34}$$

Assuming isentropic flow, $\rho/\rho_{01} = (T/T_{01})^{1/(\gamma-1)}$. Using the continuity equation,

$$\frac{\dot{m}}{A} = \rho_{01}a_{01}\left(\frac{T}{T_{01}}\right)^{(\gamma+1)/2(\gamma-1)}$$

$$= \rho_{01}a_{01}\left[\frac{2}{\gamma+1}\left(1 + \frac{U^2}{2C_pT_{01}}\right)\right]^{(\gamma+1)/2(\gamma-1)}$$

$$= \rho_{01}a_{01}\left[\frac{2 + (\gamma-1)U^2/a_{01}^2}{\gamma+1}\right]^{(\gamma+1)/2(\gamma-1)} \tag{7.35}$$

If choking occurs in the rotating passages, eqn. (7.35) indicates that the mass flow is dependent on the blade speed. As the speed of rotation is increased the compressor can accept a *greater* mass flow, unless choking occurs in some other component of the compressor. That the choking flow in an impeller can vary, depending on blade speed, may seem at first rather surprising; the above analysis gives the *reason* for the variation of the choking limit of a compressor.

Diffuser

The relation for the choking flow, eqn. (7.33) holds for the diffuser passages, it being noted that stagnation conditions now refer to the diffuser and not the inlet. Thus

$$\frac{\dot{m}}{A_2} = \rho_{02}a_{02}\left(\frac{2}{\gamma+1}\right)^{(\gamma+1)/2(\gamma-1)}. \tag{7.36}$$

Clearly, stagnation conditions at diffuser inlet are dependent on the impeller process. To find how the choking mass flow limit is affected by blade speed it is necessary to refer back to inlet stagnation conditions.
Assuming a radial bladed impeller of efficiency η_i then,

$$T_{02s} - T_{01} = \eta_i(T_{02} - T_{01}) = \eta_i\sigma U_2^2/C_p.$$

Hence

$$p_{02}/p_{01} = (T_{02s}/T_{01})^{\gamma/(\gamma-1)} = [1 + \eta_i\sigma U_2^2/(C_pT_{01})]^{\gamma/(\gamma-1)}$$

and

$$T_{02}/T_{01} = [1 + \sigma U_2^2/(C_p T_{01})].$$

Now

$$\rho_{02} a_{02} = \rho_{01} a_{01} (\rho_{02}/\rho_{01})(a_{02}/a_{01})$$
$$= \rho_{01} a_{01} [\rho_{02}/\rho_{01}(T_{01}/T_{02})^{\frac{1}{2}}],$$

therefore,

$$\frac{\dot{m}}{A_2} = \rho_{01} a_{01} \frac{[1 + (\gamma - 1)\eta_i \sigma U_2^2/a_{01}^2]^{\gamma/(\gamma-1)}}{[1 + (\gamma - 1)\sigma U_2^2/a_{01}]^{\frac{1}{2}}} \left(\frac{2}{\gamma + 1}\right)^{(\gamma+1)/2(\gamma-1)}. \tag{7.37}$$

In this analysis it should be noted that the diffuser process has been assumed to be isentropic but the impeller has not. Equation (7.37) indicates that the choking mass flow can be varied by changing the impeller rotational speed.

REFERENCES

1. CHESHIRE, L. J., The design and development of centrifugal compressors for aircraft gas turbines. *Proc. Instn. Mech. Engrs. London*, **153** (1945); reprinted by the A.S.M.E. (1947), Lectures on the development of the British gas turbine jet unit.
2. CAMPBELL, K. and TALBERT, J. E., Some advantages and limitations of centrifugal and axial aircraft compressors. *S.A.E. Journal (Transactions)*, **53**, 10 (1945).
3. MOULT, E. S. and PEARSON, H., The relative merits of centrifugal and axial compressors for aircraft gas turbines. *J. Roy. Aero. Soc.* **55** (1951).
4. DEAN, R. C., Jr., The centrifugal compressor. *Creare Inc. Technical Note* TN183 (1973).
5. SHEPHERD, D. G., *Principles of Turbomachinery*. Macmillan, New York (1956).
6. STANITZ, J. D., Some theoretical aerodynamic investigations of impellers in radial and mixed flow centrifugal compressors. *Trans. A.S.M.E.* **74**, 4 (1952).
7. STODOLA, A., *Steam and Gas Turbines*. Vols. I and II. McGraw-Hill, New York (1927). (Reprinted, Peter Smith, New York (1945).)
8. WISLICENUS, G. F., *Fluid Mechanics of Turbomachinery*. McGraw-Hill, New York (1947).
9. FERGUSON, T. B., *The Centrifugal Compressor Stage*. Butterworth, London (1963).
10. CSANADY, G. T., Head correction factors for radial impellers. *Engineering, London*, **190** (1960).
11. PEARSALL, I. S., *Cavitation*. M&B Monograph ME/10. Mills & Boon (1972).
12. SCHORR, P. G., WELLIVER, A. D. and WINSLOW, L. J., Design and development of small, high pressure ratio, single stage centrifugal compressors. *Advanced Centrifugal Compressors. Am. Soc. Mech. Engrs.* (1971).

PROBLEMS

NOTE. In problems 1 to 5 assume γ and R are 1·4 and 287 J/(kg°C) respectively. In problems 1 to 4 assume the stagnation pressure and stagnation temperature at compressor entry are 101·3 kPa and 288 K respectively.)

1. The air entering the impeller of a centrifugal compressor has an absolute axial velocity of 100 m/s. At rotor exit the relative air angle measured from the radial direction is 26° 36′, the radial component of velocity is 120 m/s and the tip speed of the radial vanes is 500 m/s. Determine the power required to drive the compressor when the air flow rate is 2·5 kg/s and the mechanical efficiency is 95%. If the radius ratio of the impeller eye is 0·3, calculate a suitable inlet diameter assuming the inlet flow is incompressible. Determine the overall total pressure ratio of the compressor when the total-to-total efficiency is 80%, assuming the velocity at exit from the diffuser is negligible.

2. A centrifugal compressor has an impeller tip speed of 366 m/s. Determine the absolute Mach number of the flow leaving the radial vanes of the impeller when the radial component of velocity at impeller exit is 30·5 m/s and the slip factor is 0·90. Given that the flow area at impeller exit is 0·1 m² and the total-to-total efficiency of the impeller is 90%, determine the mass flow rate.

3. The eye of a centrifugal compressor has a hub/tip radius ratio of 0·4, a maximum relative flow Mach number of 0·9 and an absolute flow which is uniform and completely axial. Determine the optimum speed of rotation for the condition of maximum mass flow given that the mass flow rate is 4·536 kg/s. Also, determine the outside diameter of the eye and the ratio of axial velocity/blade speed at the eye tip. Figure 7.4 may be used to assist the calculations.

4. An experimental centrifugal compressor is fitted with free-vortex guide vanes in order to reduce the relative air speed at inlet to the impeller. At the outer radius of the eye, air leaving the guide-vanes has a velocity of 91·5 m/s at 20 deg to the axial direction. Determine the inlet relative Mach number, assuming frictionless flow through the guide vanes, and the impeller total-to-total efficiency.

Other details of the compressor and its operating conditions are:

Impeller entry tip diameter, 0·457 m
Impeller exit tip diameter, 0·762 m
Slip factor 0·9
Radial component of velocity at impeller exit, 53·4 m/s
Rotational speed of impeller, 11,000 rev/min
Static pressure at impeller exit, 223 kPa (abs.)

5. A centrifugal compressor has an impeller with 21 vanes, which are radial at exit, a vaneless diffuser and no inlet guide vanes. At inlet the stagnation pressure is 100 kPa abs. and the stagnation temperature is 300 K.

(i) Given that the mass flow rate is 2·3 kg/s, the impeller tip speed is 500 m/s and the mechanical efficiency is 96%, determine the driving power on the shaft. Use eqn. (7.18a) for the slip factor.

(ii) Determine the total and static pressures at diffuser exit when the velocity at that position is 100 m/s. The total to total efficiency is 82%.

(iii) The reaction, which may be defined as for an axial flow compressor by eqn. (5.10b), is 0·5, the absolute flow speed at impeller entry is 150 m/s and the diffuser efficiency is 84%. Determine the total and static pressures, absolute Mach number and radial component of velocity at the impeller exit.

(iv) Determine the total-to-total efficiency for the impeller.

(v) Estimate the inlet/outlet radius ratio for the diffuser assuming the conservation of angular momentum.

(vi) Find a suitable rotational speed for the impeller given an impeller tip width of 6 mm.

6. A centrifugal pump is used to raise water against a static head of 18·0 m. The suction and delivery pipes, both 0·15 m diameter, have respectively, friction head losses amounting to 2·25 and 7·5 times the dynamic head. The impeller, which rotates at 1450 rev/min, is 0·25 m diameter with 8 vanes, radius ratio 0·45, inclined backwards at $\beta_2' = 60$ deg. The axial width of the impeller is designed so as to give constant radial velocity at all radii and is 20 mm at impeller exit. Assuming an hydraulic efficiency of 0·82 and an overall efficiency of 0·72, determine

(i) the volume flow rate;

(ii) the slip factor using Busemann's method;

(iii) the impeller vane inlet angle required for zero incidence angle;

(iv) the power required to drive the pump.

CHAPTER 8

Radial Flow Turbines

I like work; it fascinates me, I can sit and look at it for hours.
(JEROME K. JEROME, *Three Men in a Boat*.)

INTRODUCTION

The radial flow turbine has had a long history of development being first conceived for the purpose of producing hydraulic power over 150 years ago. A French engineer, Fourneyron, developed the first commercially successful hydraulic turbine (*c.* 1830) and this was of the *radial-outflow* type. A *radial-inflow* type of hydraulic turbine was built by Francis and Boyden in the U.S.A. (*c.* 1847) which gave excellent results and was highly regarded. This type of machine is now known as the *Francis turbine*, a simplified arrangement of it being shown in Fig. 1.1. It will be observed that the flow path followed is from the radial direction to what is substantially an axial direction. A flow path in the reverse direction (radial-outflow), for a single stage turbine anyway, creates several problems one of which (discussed later) is low specific work. However, as pointed out by Shepherd,[1] radial-outflow steam turbines comprising many stages have received considerable acceptance in Europe. Figure 8.1., from Ref, 2, shows diagrammatically the *Ljungström steam turbine* which, because of the tremendous increase in specific volume of steam, makes the radial-outflow flow path virtually imperative. A unique feature of the Ljungström turbine is that it does not have any stationary blade rows. The two rows of blades comprising each of the stages rotate in opposite directions so that they can both be regarded as rotors.

The inward-flow radial (IFR) turbine covers tremendous ranges of power, rates of mass flow and rotational speeds, from very large

221

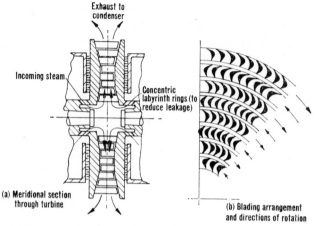

Fig. 8.1. Ljungström type outward flow radial turbine (adapted from Kearton[2])

Francis turbines used in hydroelectric power generation and developing hundreds of megawatts down to tiny closed cycle gas turbines for space power generation of a few kilowatts. Francis turbines producing 250 MW have been built in the U.S.S.R. and, according to Puyo,[3] projected designs in N. America and Russia for 500 MW or more have been considered. Further development in single unit capacity appears to be limited only by production techniques and transportation problems.

It is interesting to find that several rival claims have recently been made in different countries for the largest capacity power plant in the world all based, incidentally, upon Francis turbines. Shmelev[4] gives a technical description of a hydroelectric project at Bratsk in Eastern Siberia claimed to be the largest in the world at 4,500 MW with 20 generating sets driven by Francis turbines. Baptist and Nitta[5] give some details of the extremely large Francis turbines soon to be operational at Grand Coulee, Colorado in the U.S.A. Eventually there will be installed 12 units each of 600 MW capacity giving a record "nameplate" capacity of 7,200 MW by 1977. The rotors of these turbines are 10 m diameter, rotating at 72 rev/min and each passing water at a rate of 850 m³/s. This type of turbine normally operates with heads (of water) in the range 30–500 m. The efficiency of large Francis turbines has gradually risen over the years and now is about 95%. An historical review of this progress is given by Danel.[6] There seems to be little prospect of much further improvement in efficiency as skin friction, tip leakage and exit kinetic energy from the diffuser apparently accounts for the remaining losses.

During the last few decades the small IFR turbine, or centripetal turbine, has been used in many applications. These include automotive and diesel engine supercharging, expansion units in aircraft cooling systems, expansion units in gas liquefaction and other cryogenic systems and as a component of the small gas turbines used for space power generation.[7] Over a limited range of specific speed they provide an efficiency about equal to that of the best axial flow turbines. The IFR turbine also has the advantage of ease of manufacture and sturdy construction.

TYPES OF INWARD FLOW RADIAL TURBINE

In the centripetal turbine energy is transferred from the fluid to the rotor in passing from a large radius to a small radius. For the production of positive work the product of Uc_θ at entry to the rotor must be greater than Uc_θ at rotor exit (eqn. (2.12b)). This is usually arranged by imparting a large component of tangential velocity at rotor entry, using single or multiple nozzles, and allowing little or no swirl in the exit absolute flow.

Cantilever turbine

Figure 8.2a shows a *cantilever* IFR turbine where the blades are limited to the region of the rotor tip, extending from the rotor in the *axial* direction. In practice the cantilever blades are usually of the impulse type (i.e. low reaction), by which it is implied that there is little change in relative velocity at inlet and outlet of the rotor. There is no fundamental reason why the blading should not be of the reaction type. However, the resulting expansion through the rotor would require an increase in flow area. This extra flow area is extremely difficult to accommodate in a small radial distance, especially as the radius decreases through the rotor row.

Aerodynamically, the cantilever turbine is similar to an axial-impulse turbine and can even be designed by similar methods. Figure 8.2b shows the velocity triangles at rotor inlet and outlet. The fact that the flow is radially inwards hardly alters the design procedure because the blade radius ratio r_2/r_3 is close to unity anyway.

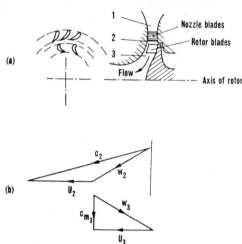

FIG. 8.2. Arrangement of cantilever turbine and velocity triangles at the design point.

The 90 degree IFR turbine

Because of its higher structural strength compared with the cantilever turbine, the 90 deg IFR turbine is the preferred type for use in turbo-chargers, gas turbines and cryogenic expanders, etc. Figure 8.3 shows the configuration and velocity triangles of a 90 deg IFR turbine; the rotor blades extend from a radially inward inlet to an axial outlet. The exit part of the blades is curved to remove most (if not all of) the absolute tangential component of velocity. This curved section of blading is known as the *exducer*. The 90 deg IFR turbine or centripetal turbine is very similar in appearance to the centrifugal compressor of Chapter 7 but with the flow direction and blade motion reversed.

The fluid discharging from the turbine rotor may have a considerable velocity c_3 and an axial diffuser would then be considered as a means of

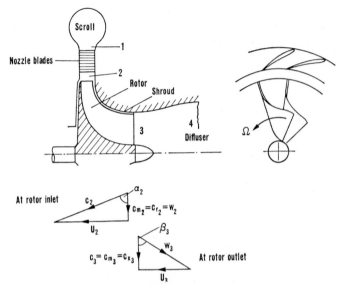

FIG. 8.3. Layout and velocity diagrams for a 90 deg inward flow radial turbine at the nominal design point.

recovering most of the kinetic energy, $\frac{1}{2}c_3{}^2$, which would otherwise be wasted. In hydraulic turbines a diffuser is invariably used and is known as the *draught tube*.

THERMODYNAMICS OF THE 90 DEG IFR TURBINE

The complete adiabatic expansion process for a turbine comprising a nozzle blade row, a radial rotor followed by a diffuser corresponding to the layout of Fig. 8.3, is represented by the Mollier diagram shown in Fig. 8.4. In the turbine, frictional processes cause the entropy to increase in all components and these irreversibilities are implied in Fig. 8.4.

Across the nozzle blades the stagnation enthalpy is constant, $h_{01} = h_{02}$ and, therefore, the static enthalpy drop is,

$$h_1 - h_2 = \tfrac{1}{2}(c_2{}^2 - c_1{}^2) \tag{8.1}$$

corresponding to the static pressure change from p_1 to the lower

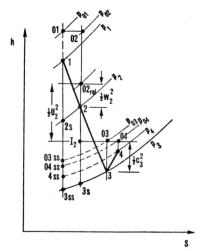

Fig. 8.4. Mollier diagram for a 90 deg inward flow radial turbine and diffuser (at the design point).

pressure p_2. The *ideal* static enthalpy change $(h_1 - h_{2s})$ is between the *same* two pressures.

In Chapter 7 it was shown that the rothalpy, $I = h_{0\,\mathrm{rel}} - \tfrac{1}{2}U^2$, is constant for an adiabatic irreversible flow process, relative to a rotating component. For the rotor of the 90 deg IFR turbine,

$$h_{02\,\mathrm{rel}} - \tfrac{1}{2}U_2^{\,2} = h_{03\,\mathrm{rel}} - \tfrac{1}{2}U_3^{\,2}$$

Thus, as $h_{0\,\mathrm{rel}} = h + \tfrac{1}{2}w^2$,

$$h_2 - h_3 = \tfrac{1}{2}[(U_2^{\,2} - U_3^{\,2}) - (w_2^{\,2} - w_3^{\,2})] \qquad (8.2)$$

In this analysis the reference point 2 (Fig. 8.3) is taken to be at the inlet radius r_2 of the rotor (the blade tip speed $U_2 = \Omega r_2$). This implies that the nozzle irreversibilities are lumped together with any friction losses occurring in the annular space between nozzle exit and rotor entry (usually scroll losses are included as well).

Across the diffuser the stagnation enthalpy does not change, $h_{03} = h_{04}$, but the static enthalpy *increases* as a result of the velocity diffusion. Hence,

$$h_4 - h_3 = \tfrac{1}{2}(c_3^{\,2} - c_4^{\,2}) \qquad (8.3)$$

The specific work done by the fluid on the rotor is

$$\Delta W = h_{01} - h_{03} = U_2 c_{\theta 2} - U_3 c_{\theta 3} \tag{8.4}$$

As $h_{01} = h_{02}$,

$$\Delta W = h_{02} - h_{03} = h_2 - h_3 + \tfrac{1}{2}(c_2^2 - c_3^2)$$
$$= \tfrac{1}{2}[(U_2^2 - U_3^2) - (w_2^2 - w_3^2) + (c_2^2 - c_3^2)] \tag{8.4a}$$

after substituting eqn. (8.2). It is the large positive contribution of the first term of eqn. (8.4a), viz. $\tfrac{1}{2}(U_2^2 - U_3^2)$, to the specific work which confers upon the inward radial flow turbine its main advantage over the outward radial flow type.

The "nominal" design condition for the 90 deg IFR turbine is defined here as a zero incidence relative flow at rotor inlet (i.e. a *radial* relative velocity, $w_2 = c_{r2}$) and an *axial* absolute velocity at rotor exit ($c_3 = c_{x3}$). Any variation from these flow conditions will be regarded, for the purposes of this analysis, as being an "off-design" condition. Small amounts of swirl at exit may actually be beneficial to flow stability in the diffuser and slightly non-radial relative flow at rotor inlet can improve efficiency, but these considerations are not taken into account at this stage. Thus, with $c_{\theta 3} = 0$ and $c_{\theta 2} = U_2$, the specific work at the design point is simply

$$\Delta W = U_2^2 \tag{8.4b}$$

At the design condition, referring to the velocity triangles of Fig. 8.3, $w_3^2 - U_3^2 = c_3^2$, and so eqn. (8.2) can be rewritten as

$$h_2 - h_3 = \tfrac{1}{2}(U_2^2 - w_2^2 + c_3^2) \tag{8.2a}$$

This particular relationship, in the form $I_2 = h_{02\ \text{rel}} - \tfrac{1}{2}U_2^2 = h_{03}$ can be easily identified in Fig. 8.4.

Again, referring to the velocity triangles, $w_2 = U_2 \cot \alpha_2$ and $c_3 = U_3 \cot \beta_3$, a useful alternative form to eqn. (8.2a) is obtained,

$$h_2 - h_3 = \tfrac{1}{2}U_2^2[(1 - \cot^2 \alpha_2) + (r_3/r_2)^2 \cot^2 \beta_3] \tag{8.2b}$$

where U_3 is written as $U_2 r_3/r_2$. For a perfect gas the temperature ratio T_3/T_2 can be easily found. Substituting $h = C_p T = \gamma R T/(\gamma - 1)$ in eqn. (8.2b)

$$1 - \frac{T_3}{T_2} = \tfrac{1}{2} U_2^2 \frac{(\gamma - 1)}{\gamma R T_2}\left[1 - \cot^2 \alpha_2 + \left(\frac{r_3}{r_2}\right)^2 \cot^2 \beta_3\right]$$

$$\therefore \quad \frac{T_3}{T_2} = 1 - \tfrac{1}{2}(\gamma - 1)\left(\frac{U_2}{a_2}\right)^2 \left[1 - \cot^2 \alpha_2 + \left(\frac{r_3}{r_2}\right)^2 \cot^2 \beta_3\right]$$

$$(8.2c)$$

where $a_2 = (\gamma R T_2)^{\frac{1}{2}}$ is the sonic velocity at temperature T_2.

The term *spouting velocity* c_0 (originating from hydraulic turbine practice) is defined as that velocity which has an associated kinetic energy equal to the isentropic enthalpy drop from turbine inlet stagnation pressure p_{01} to the final exhaust pressure. The exhaust pressure here can have several interpretations depending upon whether total or static conditions are used in the related efficiency definition and upon whether or not a diffuser is included with the turbine. Thus, when *no* diffuser is used

$$\tfrac{1}{2}c_0{}^2 = h_{01} - h_{03ss} \qquad (8.5a)$$

or,

$$\tfrac{1}{2}c_0{}^2 = h_{01} - h_{3ss} \qquad (8.5b)$$

for the total and static cases respectively.

In an *ideal* (frictionless) radial turbine with complete recovery of the exhaust kinetic energy,

$$\Delta W = U_2{}^2 = \tfrac{1}{2}c_0{}^2$$

$$\frac{U_2}{c_0} = 0.707$$

At the best efficiency point of actual (frictional) 90 deg IFR turbines it is found that this velocity ratio is, generally, in the range $0.68 < U_2/c_0 < 0.71$.

DESIGN POINT EFFICIENCY

Referring to Fig. 8.4, the total-to-static efficiency in the absence of a diffuser, is defined as

$$\eta_{ts} = \frac{h_{01} - h_{03}}{h_{01} - h_{3ss}} = \frac{\Delta W}{\Delta W + \tfrac{1}{2}c_3{}^2 + (h_3 - h_{3s}) + (h_{3s} - h_{3ss})}$$

$$(8.6)$$

The passage enthalpy losses can be expressed as a fraction (ζ) of the exit kinetic energy relative to the nozzle row and the rotor, i.e.

$$h_3 - h_{3s} = \tfrac{1}{2}w_3{}^2\zeta_R \qquad (8.7a)$$

$$h_{3s} - h_{3ss} = \tfrac{1}{2}c_2{}^2\zeta_N(T_3/T_2) \qquad (8.7b)$$

for the rotor and nozzles respectively. It is noted that for a constant pressure process, $Tds = dh_2$, hence the approximation,

$$h_{3s} - h_{3ss} = (h_2 - h_{2s})(T_3/T_2)$$

Substituting for the enthalpy losses in eqn. (8.6),

$$\eta_{ts} = [1 + \tfrac{1}{2}(c_3{}^2 + w_3{}^3\zeta_R + c_2{}^2\zeta_N T_3/T_2)/\Delta W]^{-1} \qquad (8.8)$$

From the design point velocity triangles, Fig. 8.3,

$$c_2 = U_2 \operatorname{cosec} a_2, \; w_3 = U_3 \operatorname{cosec} \beta_3, \; c_3 = U_3 \cot \beta_3, \; \Delta W = U_2{}^2.$$

Thus, substituting all these expressions in eqn. (8.8) and noting that $U_3 = U_2 r_3/r_2$, then

$$\eta_{ts} = \left[1 + \tfrac{1}{2}\left\{\zeta_N \frac{T_3}{T_2} \operatorname{cosec}^2 a_2 + \left(\frac{r_3}{r_2}\right)^2 (\zeta_R \operatorname{cosec}^2 \beta_3 + \cot^2 \beta_3)\right\}\right]^{-1}$$

$$(8.9)$$

Usually r_3 and β_3 are taken to apply at the arithmetic mean radius, i.e. $r_3 = \tfrac{1}{2}(r_{3t} + r_{3h})$. The temperature ratio (T_3/T_2) in eqn. (8.9) can be obtained from the thermodynamic relation, eqn. (8.2c). Thus,

$$\frac{T_3}{T_2} = 1 - \tfrac{1}{2}(\gamma - 1) \left(\frac{U_2}{a_2}\right)^2 \left\{ 1 - \cot^2 a_2 + \left(\frac{r_3}{r_2}\right)^2 \cot^2 \beta_3 \right\}$$

but, generally, this temperature ratio will only have a very minor effect upon the numerical value of η_{ts} and so it is usually ignored in calculations. Thus,

$$\eta_{ts} \simeq \left[1 + \tfrac{1}{2}\left\{\zeta_N \cosec^2 \alpha_2 + \left(\frac{r_{3av}}{r_2}\right)^2 \left(\zeta_R \cosec^2 \beta_{3av} + \cot^2 \beta_{3av}\right)\right\}\right]^{-1}$$

$$(8.9a)$$

is the expression normally used to determine the total-to-static efficiency. An alternative form for η_{ts} can be obtained by rewriting eqn. (8.6) as

$$\eta_{ts} = \frac{h_{01} - h_{03}}{h_{01} - h_{3ss}} = \frac{(h_{01} - h_{3ss}) - (h_{03} - h_3) - (h_3 - h_{3s}) - (h_{3s} - h_{3ss})}{(h_{01} - h_{3ss})}$$

$$= 1 - (c_3^2 + \zeta_N c_2^2 + \zeta_R w_3^2)/c_0^2 \qquad (8.10)$$

where the spouting velocity c_0 is defined by,

$$h_{01} - h_{3ss} = \tfrac{1}{2}c_0^2 = C_p T_{01}[1 - (p_3/p_{01})^{(\gamma-1)/\gamma}] \qquad (8.11)$$

A simple connection exists between total-to-total and total-to-static efficiency which can be obtained as follows. Writing

$$\Delta W = U_2^2 = \eta_{ts}\Delta W_{is} = \eta_{ts}(h_{01} - h_{3ss})$$

then,

$$\eta_{tt} = \frac{\Delta W}{\Delta W_{is} - \tfrac{1}{2}c_3^2} = \frac{1}{\dfrac{1}{\eta_{ts}} - \dfrac{c_3^2}{2\,\Delta W}}$$

$$\therefore \frac{1}{\eta_{tt}} = \frac{1}{\eta_{ts}} - \frac{c_3^2}{2\Delta W}$$

$$= \frac{1}{\eta_{ts}} - \frac{1}{2}\left(\frac{r_{3av}}{r_2}\cot \beta_{3av}\right)^2 \qquad (8.12)$$

EXAMPLE: Performance data from the CAV type 01 radial turbine (Benson *et al.*, reference 8) operating at a pressure ratio p_{01}/p_3 of 1·5 with zero incidence relative flow onto the rotor, is presented in the following form:

$$\dot{m}\sqrt{T_{01}}/p_{01} = 1\cdot44 \times 10^{-5}, \text{ ms(deg.K)}^{\tfrac{1}{2}}$$

$$N/\sqrt{T_{01}} = 2410, \text{ (rev/min)/(deg. K)}^{\tfrac{1}{2}}$$

$$\tau/p_{01} = 4\cdot59 \times 10^{-6}, \text{ m}^3$$

where τ is the torque, corrected for bearing friction loss. The principal dimensions and angles, etc. are given as follows:

Rotor inlet diameter,	72·5 mm
Rotor inlet width,	7·14 mm
Rotor mean outlet diameter,	34·4 mm
Rotor outlet annulus width,	20·1 mm
Rotor inlet angle,	0 deg
Rotor outlet angle,	53 deg
Number of rotor blades,	10
Nozzle outlet diameter,	74·1 mm
Nozzle outlet angle,	80 deg
Nozzle blade number,	15

The turbine is "cold tested" with air heated to 400 K to prevent condensation erosion of the blades. At nozzle outlet an estimate of the flow angle is given as 71 deg and the corresponding enthalpy loss coefficient is stated to be 0·065. Assuming that the absolute flow at rotor exit is without swirl and uniform, and the relative flow leaves the rotor without any deviation, determine the total-to-static and overall efficiencies of the turbine, the rotor enthalpy coefficient and the rotor relative velocity ratio.

Solution:

The data given is obtained from an actual turbine test and, even though the bearing friction loss has been corrected, there is an additional reduction in the specific work delivered due to disk friction and tip leakage losses, etc. The rotor speed $N = 2410\sqrt{400} = 48\,200$ rev/min, the rotor tip speed $U_2 = \pi N D_2/60 = 183$ m/s and hence the specific work done by the rotor $\Delta W = U_2{}^2 = 33·48$ kJ/kg. The corresponding isentropic total-to-static enthalpy drop is

$$h_{01} - h_{3ss} = C_p T_{01}[1 - (p_3/p_{01})^{(\gamma - 1)/\gamma}]$$

$$= 1·005 \times 400[1 - (1/1·5)^{1/3·5}] = 43·97 \text{ kJ/kg}$$

Thus, the total-to-static efficiency is

$$\eta_{ts} = \Delta W/(h_{01} - h_{3ss}) = 76·14\%$$

The actual specific work output to the shaft, after allowing for the bearing friction loss, is

$$\Delta W_{act} = \tau \Omega/\dot{m} = \left(\frac{\tau}{p_{01}}\right) \frac{N}{\sqrt{T_{01}}} \left(\frac{p_{01}}{\dot{m}\sqrt{T_{01}}}\right) \frac{\pi}{30} T_{01}$$

$$= 4.59 \times 10^{-6} \times 2410 \times \pi \times 400/(30 \times 1.44 \times 10^{-5})$$

$$= 32.18 \text{ kJ/kg}$$

Thus, the turbine overall total-to-static efficiency is

$$\eta_0 = \Delta W_{act}/(h_{01} - h_{3ss}) = 73.18\%$$

By rearranging eqn. (8.9a) the rotor enthalpy loss coefficient can be obtained:

$$\zeta_R = \{2(1/\eta_{ts} - 1) - \zeta_N \operatorname{cosec}^2 \alpha_2\}(r_2/r_{3av})^2 \sin^2 \beta_{3av} - \cos^2 \beta_{3av}$$

$$= \{2(1/0.7613 - 1) - 0.065 \times 1.1186\} \times 4.442 \times 0.6378$$

$$- 0.3622$$

$$= 1.208$$

At rotor exit c_3 is assumed to be uniform and axial. From the velocity triangles, Fig. 8.4,

$$c_3 = U_3 \cot \beta_3 = U_{3av} \cot \beta_{3av} = \text{constant}$$

$$w_3^2 = U_3^2 + c_3^2$$

$$= U_{3av}^2 \left[\left(\frac{r_3}{r_{3av}}\right)^2 + \cot^2 \beta_{3av}\right]$$

$$w_{2av} = U_2 \cot \alpha_2$$

ignoring blade to blade velocity variations. Hence,

$$\frac{w_3}{w_{2av}} = \frac{r_{3av}}{r_2} \tan \alpha_2 \left[\left(\frac{r_3}{r_{3av}}\right)^2 + \cot^2 \beta_{3av}\right]^{\frac{1}{2}}. \qquad (8.13)$$

The lowest value of this relative velocity ratio occurs when r_3 is least, i.e. $r_3 = r_{3h} = (34.4 - 20.1)/2 = 7.15$ mm, so that

$$\left(\frac{w_3}{w_{2av}}\right)_{min} = 0.475 \times 2.904[0.415^2 + 0.7536^2]^{\frac{1}{2}} = 1.19.$$

The relative velocity ratio corresponding to the mean exit radius is,

$$\frac{w_{3av}}{w_{2av}} = 0.475 \times 2.904[1 + 0.7536^2]^{\frac{1}{2}} = 1.73.$$

It is worth commenting that higher total-to-static efficiencies have been obtained in other small radial turbines operating at higher pressure ratios. Rodgers[9] has suggested that total-to-static efficiencies in excess of 90% for pressure ratios up to five to one can be attained. Nusbaum and Kofskey[10] reported an experimental value of 88·8% for a small radial turbine (fitted with an outlet diffuser, admittedly!) at a pressure ratio p_{01}/p_4 of 1·763. In the design point exercise given above the high rotor enthalpy loss coefficient and the corresponding relatively low total-to-static efficiency may well be related to the low relative velocity ratio determined on the hub. Matters are probably worse than this as the calculation is based only on a simple one-dimensional treatment. In determining velocity ratios across the rotor, account should also be taken of the effect of blade to blade velocity variation (outlined in this chapter) as well as viscous effects. The number of vanes in the impeller (ten) may be insufficient on the basis of Jamieson's theory* (reference 14) which suggests 18 vanes (i.e. $Z_{min} = 2\pi \tan a_2$). For this turbine, at lower nozzle exit angles, eqn. (8.13) suggests that the relative velocity ratio becomes even less favourable despite the fact that the Jamieson blade spacing criterion is being approached. (For $Z = 10$, the optimum value of a_2 is about 58 deg.)

MACH NUMBER RELATIONS

Assuming the fluid is a perfect gas, expressions can be deduced for the important Mach numbers in the turbine. At nozzle outlet the absolute Mach number at the design point is,

$$M_2 = \frac{c_2}{a_2} = \frac{U_2}{a_2} \operatorname{cosec} a_2.$$

Now, $T_2 = T_{01} - c_2^2/(2C_p) = T_{01} - \frac{1}{2}U_2^2 \operatorname{cosec}^2 a_2/C_p.$

$$\therefore \quad \frac{T_2}{T_{01}} = 1 - \frac{1}{2}(\gamma - 1)(U_2/a_{01})^2 \operatorname{cosec}^2 a_2$$

where $a_2 = a_{01}(T_2/T_{01})^{\frac{1}{2}}$. Hence,

$$M_2 = \frac{U_2/a_{01}}{\sin a_2[1 - \frac{1}{2}(\gamma - 1)(U_2/a_{01})^2 \operatorname{cosec}^2 a_2]^{\frac{1}{2}}} \tag{8.14}$$

* Included in a later part of this Chapter.

At impeller outlet the relative Mach number at the design point is defined by,

$$M_{r3} = \frac{w_3}{a_3} = \frac{r_3 U_2}{r_2 a_3} \operatorname{cosec} \beta_3.$$

Now,

$$h_3 = h_{01} - (U_2{}^2 + \tfrac{1}{2}c_3{}^2) = h_{01} - (U_2{}^2 + \tfrac{1}{2}U_3{}^2 \cot^2 \beta_3)$$

$$= h_{01} - U_2{}^2 \left[1 + \tfrac{1}{2} \left(\frac{r_3}{r_2} \cot \beta_3 \right)^2 \right]$$

$$a_3{}^2 = a_{01}{}^2 - (\gamma - 1)U_2{}^2 \left[1 + \tfrac{1}{2} \left(\frac{r_3}{r_2} \cot \beta_3 \right)^2 \right]$$

$$\therefore M_{r3} = \frac{(U_2/a_{01})(r_3/r_2)}{\sin \beta_3 \left[1 - (\gamma - 1)(U_2/a_{01})^2 \left\{ 1 + \tfrac{1}{2} \left(\frac{r_3}{r_2} \cot \beta_3 \right)^2 \right\} \right]^{\frac{1}{2}}}$$

$$(8.15)$$

LOSS COEFFICIENTS IN 90 DEG IFR TURBINES

There are a number of ways of representing the losses in the passages of 90 deg IFR turbines and these have been listed and inter-related by Benson.[11] As well as the nozzle and rotor passage losses there is, in addition, a loss at rotor entry at off-design conditions. This occurs when the relative flow entering the rotor is at some angle of incidence to the radial vanes so that it can be called an *incidence loss*. It is often referred to as a "shock loss" but this can be rather misleading because, usually, there is no shock wave.

(i) *Nozzle loss coefficients*

The enthalpy loss coefficient, which normally includes the inlet scroll losses, has already been defined and is,

$$\zeta_N = (h_2 - h_{2s})/(\tfrac{1}{2}c_2{}^2). \tag{8.16}$$

Also in use is the *velocity coefficient*,

$$\phi_N = c_2/c_{2s} \qquad (8.17)$$

and the *stagnation pressure loss coefficient*,

$$Y_N = (p_{01} - p_{02})/(p_{02} - p_2) \qquad (8.18a)$$

which can be related, approximately, to ζ_N by

$$Y_N \simeq \zeta_N(1 + \tfrac{1}{2}\gamma M_2^{\,2})\zeta_N \qquad (8.18b)$$

Since, $h_{01} = h_2 + \tfrac{1}{2}c_2^{\,2} = h_{2s} + \tfrac{1}{2}c_{2s}^{\,2}$, then $h_2 - h_{2s} = \tfrac{1}{2}(c_{2s}^{\,2} - c_2^{\,2})$
and

$$\zeta_N = \frac{1}{\phi_N^{\,2}} - 1. \qquad (8.19)$$

Practical values of ϕ_N for well-designed nozzle rows in normal operation are usually in the range $0{\cdot}90 \leqslant \phi_N \leqslant 0{\cdot}97$.

(ii) *Rotor loss coefficients*

At either the design condition (Fig. 8.4), or at the off-design condition dealt with later (Fig. 8.5), the rotor passage friction losses can be expressed in terms of the following coefficients.

The enthalpy loss coefficient is,

$$\zeta_R = (h_3 - h_{3s})/(\tfrac{1}{2}w_3^{\,2}). \qquad (8.20)$$

The velocity coefficient is,

$$\phi_R = w_3/w_{3s} \qquad (8.21)$$

which is related to ζ_R by

$$\zeta_R = \frac{1}{\phi_R^{\,2}} - 1 \qquad (8.22)$$

The normal range of ϕ for well-designed rotors is approximately, $0{\cdot}75 \leqslant \phi_R \leqslant 0{\cdot}85$.

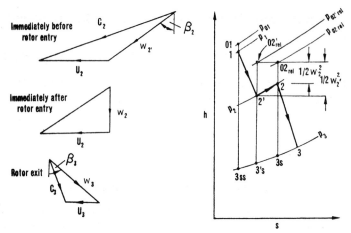

FIG. 8.5. Velocity triangles and Mollier diagram for a 90 deg IFR turbine at an
off-design condition.

OFF-DESIGN OPERATING CONDITION

At off-design conditions of operation (e.g. when the speed of rotation
is lower than design speed), an additional loss due to flow incidence on
to the rotor can be identified. Futral and Wasserbauer[12] defined this
loss as equal to the kinetic energy corresponding to the component of
velocity normal to the rotor vane at inlet. Immediately *before* entering
the rotor the relative velocity (Fig. 8.5) is $w_{2'}$ and this is reduced to w_2
immediately *after* entry to the rotor. Apparently, the origin of the
"shock loss" concept is owed to this simple model of a *sudden* change in
relative velocity at rotor inlet. Thus, as the radius does not change,
$h_{0\ rel}$ is constant for this process and $h_{2'} + \frac{1}{2}w_{2'}^2 = h_2 + \frac{1}{2}w_2^2$ and so

$$h_2 - h_{2'} = \frac{1}{2}(w_{2'}^2 - w_2^2) = \frac{1}{2}w_2^2 \tan^2 \beta_{2'}. \qquad (8.23)$$

This enthalpy increase is shown in the Mollier diagram (Fig. 8.5) as a
constant pressure process ($p_2 =$ constant) and there must then be a
corresponding loss in stagnation pressure equal to $(p_{02'rel} - p_{02rel}) = (\Delta p_0)_{sh}$. An *incompressible* flow definition for the incidence loss coefficient is sometimes convenient and is

$$Y_{sh} = (\Delta p_0)_{sh}/(\tfrac{1}{2}\rho w_2^2) = \tan^2 \beta_{2'}. \qquad (8.24a)$$

For *compressible* flow, the incidence loss coefficient has the equivalent form,

$$Y_{sh} = (\Delta p_0)_{sh}/[(p_{02'rel} - p_2) \cos^2 \beta_{2'}] = \tan^2 \beta_{2'} \quad (8.24b)$$

Bridle and Boulter[13] quote an empirical result for Y_{sh} which closely fits test data in the range $\beta_{2'} = \pm 65$ deg, with the expression

$$Y_{sh} = (\tan \beta_{2'} + 0.1)^2. \quad (8.24c)$$

CRITERION FOR MINIMUM NUMBER OF BLADES

The following simple analysis of the relative flow in a radially bladed rotor is of considerable interest as it illustrates an important fundamental point concerning blade spacing. From elementary mech-

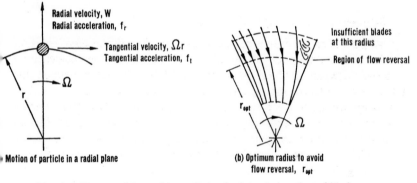

Radial velocity, W
Radial acceleration, f_r

Tangential velocity, Ωr
Tangential acceleration, f_t

Insufficient blades at this radius

Region of flow reversal

r_{opt}

Motion of particle in a radial plane

(b) Optimum radius to avoid flow reversal, r_{opt}

FIG. 8.6. Flow models used in analysis of minimum number of blades.

anics, the radial and transverse components of acceleration, f_r and f_t respectively, of a particle moving in a radial plane (Fig. 8.6a) are:

$$f_r = \dot{w} - \Omega^2 r \quad (8.25a)$$

$$f_t = r\dot{\Omega} + 2\Omega w \quad (8.25b)$$

where w is the radial velocity, $\dot{w} = \dfrac{dw}{dt} = w\dfrac{\partial w}{\partial r}$ (for steady flow), Ω is the angular velocity and $\dot{\Omega} = d\Omega/dt$ is set equal to zero.

Applying *Newton's second law of motion* to a fluid element (as shown in Fig. 6.2) of unit depth, ignoring viscous forces, but putting $c_r = w$, the radial equation of motion is,

$$(p + dp)(r + dr)d\theta - prd\theta - pdrd\theta = -f_r dm$$

where the elementary mass $dm = \rho rd\theta dr$. After simplifying and substituting for f_r from eqn. (8.25a), the following result is obtained,

$$\frac{1}{\rho}\frac{\partial p}{\partial r} + w\frac{\partial w}{\partial r} = \Omega^2 r. \tag{8.26}$$

Integrating eqn. (8.26) with respect to r obtains

$$p/\rho + \tfrac{1}{2}w^2 - \tfrac{1}{2}U^2 = \text{constant} \tag{8.27}$$

which is merely the *inviscid form* of eqn. (8.2).

The torque transmitted to the rotor by the fluid manifests itself as a pressure difference across each radial vane. Consequently, there must be a pressure gradient in the *tangential direction* in the space between the vanes. Again, consider the element of fluid and apply Newton's second law of motion in the tangential direction

$$dp.dr = f_t dm = 2\Omega w(\rho rd\theta dr).$$

Hence,

$$\frac{1}{\rho}\frac{\partial p}{\partial \theta} = 2\Omega rw \tag{8.28}$$

which establishes the magnitude of the tangential pressure gradient.

Differentiating eqn. (8.27) with respect to θ,

$$\frac{1}{\rho}\frac{\partial p}{\partial \theta} = -w\frac{\partial w}{\partial \theta}. \tag{8.29}$$

Thus, combining eqns. (8.28) and (8.29) gives,

$$\frac{\partial w}{\partial \theta} = -2\Omega r \tag{8.30}$$

This result establishes the important fact that *the radial velocity is not uniform across the passage* as is frequently assumed. As a consequence of this fact the radial velocity on one side of a passage is lower than on the other side. Jamieson,[14] who originated this method, conceived the idea of determining the *minimum* number of blades based upon these velocity considerations.

Let the mean radial velocity be $\bar{w}$ and the angular space between two adjacent blades be $\Delta\theta = 2\pi/Z$ where Z is the number of blades. The maximum and minimum radial velocities are, therefore,

$$w_{max} = \bar{w} + \tfrac{1}{2}\Delta w = \bar{w} + \Omega r\Delta\theta \qquad (8.31\text{a})$$

$$w_{min} = \bar{w} - \tfrac{1}{2}\Delta w = \bar{w} - \Omega r\Delta\theta \qquad (8.31\text{b})$$

using eqn. (8.30).

Making the reasonable assumption that the radial velocity should not drop below zero, (see Fig. 8.6b), then the limiting case occurs at the impeller tip, $r = r_2$ with $w_{min} = 0$. From eqn. (8.31b) with $U_2 = \Omega r_2$, the minimum number of rotor blades is

$$Z_{min} = 2\pi U_2/\bar{w}_2 \qquad (8.32\text{a})$$

At the design condition, $U_2 = \bar{w}_2 \tan \alpha_2$, hence

$$Z_{min} = 2\pi \tan \alpha_2 \qquad (8.32\text{b})$$

Figure 8.7 shows the variation of Z_{min} with α_2 according to eqn. (8.32b).

A more rigorous analysis than the above treatment has been advanced by Wallace[15] which takes into account the compressible nature of the flow in a radial turbine, blade inclination and flow curvature. Wallace's analysis predicts a rather larger number of vanes to avoid flow reversal than the simpler treatment of Jamieson.

Some experimental tests reported by Hiett and Johnston[16] are of interest in connection with the analysis presented above. With a nozzle outlet angle $\alpha_2 = 77$ deg and a 12 vane rotor, a total-to-static efficiency $\eta_{ts} = 0.84$ was measured at the optimum velocity ratio U_2/c_0. For that magnitude of flow angle, eqn. (8.32b) suggests 27 vanes would be required in order to avoid reverse flow at the rotor tip. However, a second test with the number of vanes increased to 24 produced a gain in efficiency of only 1%. Hiett and Johnston suggested that the criterion for the optimum number of vanes might not simply be the avoidance of local flow reversal but might require a compromise between total pressure losses from this cause and friction losses based upon rotor and blade surface areas.

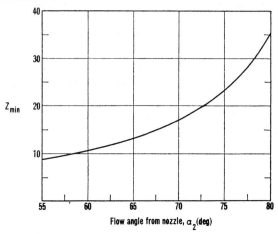

Fig. 8.7. Minimum blade number required to avoid flow reversal at rotor entry.

SIGNIFICANCE AND APPLICATION OF SPECIFIC SPEED

The concept of specific speed N_s has already been discussed in Chapter 1 and some applications of it have been made already. Specific speed is extensively used to describe turbomachinery operating requirements in terms of shaft speed, volume flow rate and ideal specific work (alternatively, power developed is used instead of specific work). Originally, specific speed was applied almost exclusively to *incompressible* flow machines as a tool in the selection of the optimum type and size of unit. Its application to units handling *compressible* fluids was somewhat inhibited, due, it would appear, to the fact that volume flow rate changes through the machine, which raised the awkward question of which flow rate should be used in the specific speed definition. According to Balje,[17] the significant volume flow rate which should be used is that in the rotor exit, Q_3. This has now been widely adopted by many authorities.

Wood[18] found it useful to factorise the basic definition of the specific speed equation, eqn. (1.8), in terms of the geometry and flow conditions within the radial-inflow turbine. Adopting the non-dimensional form of specific speed, in order to avoid ambiguities,

$$N_s = \frac{N Q_3^{\frac{1}{2}}}{\Delta h_{0s}^{\frac{3}{4}}} \qquad (8.33)$$

where N is in rev/s, Q_3 is in m³/s and the isentropic total-to-total enthalpy drop Δh_{0s} (from turbine inlet to exhaust) is in J/kg (i.e. m²/s²).

For the 90 deg IFR turbine, writing $U_2 = \pi N D_2$ and $\Delta h_{0s} = \frac{1}{2} c_0^2$, eqn. (8.33) can be factorised as follows:

$$N_s = \frac{Q_3^{\frac{1}{2}}}{(\frac{1}{2} c_0^2)^{\frac{3}{4}}} \left(\frac{U_2}{\pi D_2}\right) \left(\frac{U_2}{\pi N D_2}\right)^{\frac{1}{2}}$$

$$= \left(\frac{\sqrt{2}}{\pi}\right)^{\frac{3}{2}} \left(\frac{U_2}{c_0}\right)^{\frac{3}{2}} \left(\frac{Q_3}{N D_2^3}\right)^{\frac{1}{2}} \qquad (8.34)$$

For the *ideal* 90 deg. IFR turbine, it was shown earlier that the blade speed to spouting velocity ratio, $U_2/c_0 = 1/\sqrt{2} = 0.707$. Substituting this value into eqn. (8.34),

$$N_s = 0.18 \left(\frac{Q_3}{N D_2^3}\right)^{\frac{1}{2}}, \text{(rev)} \qquad (8.34a)$$

i.e. specific speed is directly proportional to the square root of the volumetric flow coefficient.

To obtain some physical significance from eqns. (8.33) and (8.34a), define a *rotor disc area* $A_d = \pi D_2^2/4$ and assume a uniform axial rotor exit velocity c_3 so that $Q_3 = A_3 c_3$, then as

$$N = U_2/(\pi D_2) = \frac{c_0 \sqrt{2}}{2 \pi D_2}$$

$$\frac{Q_3}{N D_2^3} = \frac{A_3 c_3 \, 2 \pi D_2}{\sqrt{2} c_0 D_2^2} = \frac{A_3}{A_d} \frac{c_3}{c_0} \frac{\pi^2}{2\sqrt{2}}$$

Hence,

$$N_s = 0.336 \left(\frac{c_3}{c_0}\right)^{\frac{1}{2}} \left(\frac{A_3}{A_d}\right)^{\frac{1}{2}}, \text{(rev)} \qquad (8.34b)$$

or,

$$\Omega_s = 2\cdot11 \left(\frac{c_3}{c_0}\right)^{\frac{1}{2}} \left(\frac{A_3}{A_d}\right)^{\frac{1}{2}}, \quad \text{(rad)} \qquad (8.34c)$$

Recent NASA design studies (e.g. ref. 7) of 90 deg IFR gas turbines suggest that the ratio of rotor exit tip diameter to rotor inlet diameter D_{3t}/D_2 should not exceed 0·7 to avoid excessive curvature of the shroud. Also, it would be unlikely that the rotor exit hub to tip diameter ratio D_{3h}/D_{3t} would fall below 0·35 because of flow blockage caused by the closely spaced blades. Thus, an upper limit for A_3/A_d can be found,

$$\frac{A_3}{A_d} = \left(\frac{D_{3t}}{D_2}\right)^2 \left[1 - \left(\frac{D_{3h}}{D_{3t}}\right)^2\right] = 0\cdot7^2 \times (1 - 0\cdot122) = 0\cdot43.$$

Fig. 8.8 shows the relationship between Ω_s, the *exhaust energy factor* $(c_3/c_0)^2$ and the area ratio A_3/A_d based upon eqn. (8.34c). According to Wood[18] the limits for the exhaust energy factor in gas turbine practice are $0\cdot04 < (c_3/c_0)^2 < 0\cdot30$, the lower value being apparently a flow stability limit.

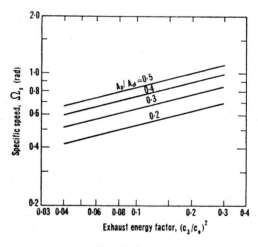

FIG. 8.8. Specific speed function for a 90 deg inward flow radial turbine (adapted from Wood[18]).

The numerical value of specific speed provides a general index of flow capacity relative to work output. Low values of Ω_s are associated with relatively small flow passage area and high values with relatively large flow passage areas. Specific speed has also been widely used as a general indication of achievable efficiency. Figure 8.9 presents a broad

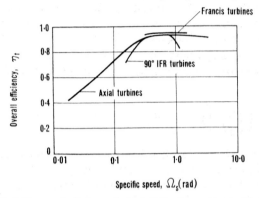

FIG. 8.9. Specific speed-efficiency characteristics for various turbines (adapted from Wood[18]).

correlation of maximum efficiencies for hydraulic and compressible fluid turbines as functions of specific speed. These efficiencies apply to favourable design conditions with high values of flow Reynolds number, efficient diffusers and low leakage losses at the blade tips. It is seen that over a limited range of specific speed the best radial-flow turbines match the best axial-flow turbine efficiency, but from $\Omega_s = 0.03$ to 10, no other form of turbine handling compressible fluids can exceed the peak performance capability of the axial turbine.

Over the fairly limited range of specific speed ($0.3 \leq \Omega_s < 0.9$) that the 90 deg IFR turbine can produce a high efficiency it is difficult to find a decisive performance advantage in favour of either the axial flow turbine or the radial-flow turbine. New methods of fabrication enable the blades of small axial-flow turbines to be cast integrally with the rotor so that both types of turbine can operate at about the same blade tip speed. Wood[18] has compared the relative merits of axial and radial gas turbines at some length. In general, although weight, bulk and diameter are greater for radial than axial turbines, the differences are not so large and mechanical design compatibility can reverse the difference in a complete gas turbine power plant. The NASA nuclear Brayton cycle space power studies have all been made with 90 deg IFR turbines rather than with axial flow turbines.

It is of interest to use the data given for the very large Francis turbines to be used at Grand Coulee (see Introduction to this chapter) to determine specific speed and to compare it with the data given in Fig. 8.9. Neither the efficiency nor the head is specifically stated. It would be reasonable to assume, however, that the efficiency, η_t, is 0·95 in order to determine the effective head at turbine entry. Thus,

$$H = \dot{W}_t/(\rho g Q \eta_t) = 600 \times 10^6/(10^3 \times 9\cdot81 \times 850 \times 0\cdot95).$$
$$= 75\cdot8\text{m}.$$

Hence, the specific speed (rad) is,

$$\Omega_s = \frac{\Omega Q^{\frac{1}{2}}}{(gH)^{\frac{3}{4}}} = \frac{(72 \times \pi/30) \times 850^{\frac{1}{2}}}{(9\cdot81 \times 75\cdot8)^{\frac{3}{4}}} \cdot$$
$$= 1\cdot56$$

It can be seen from Fig. 8.9 that this value of Ω_s is close to the right hand end of the curve marked "Francis turbines", in agreement with the general characteristic. For fixed values of Q, H and U_2 then, since $\Omega = 2U_2/D_2$,

$$\Omega_s = \frac{\Omega Q^{\frac{1}{2}}}{(gH)^{\frac{3}{4}}} \propto \frac{1}{D_2}$$

i.e. Ω_s varies inversely with diameter D_2.

Thus, given constant efficiency over a *range* of specific speed, a designer would tend to choose the highest value of Ω_s in that range as it would enable the *smallest size* turbine to be used.

OPTIMUM DESIGN SELECTION OF 90 DEG IFR TURBINES

Rohlik[19] has examined analytically the performance of 90 deg inward flow radial turbines in order to determine *optimum* design geometry for various applications as characterised by specific speed.

His procedure, which extends the treatment of Balje[17] and Wood,[18] was used to determine the design point losses and corresponding efficiencies for various combinations of nozzle exit flow angle a_2, rotor diameter ratio D_2/D_{3av} and rotor blade entry height to exit diameter ratio, b_2/D_{3av}. The losses taken into account in the calculations are those associated with,

   (i) nozzle blade row boundary layers,
   (ii) rotor passage boundary layers,
  (iii) rotor blade tip clearance,
  (iv) disc windage (on the back surface of the rotor),
   (v) kinetic energy loss at exit.

A mean-flowpath analysis was used and the passage losses were based upon the data of Stewart *et al.*[20] The main constraints in the analysis were:

(a) the rotor relative exit velocity w_{3av} was to be twice the rotor inlet relative velocity w_2 to provide a consistent rotor reaction and to be sufficiently high to assure low rotor total pressure losses,

(b) zero absolute exit whirl from the rotor,

(c) minimum loss condition at rotor inlet (zero incidence),

(d) the ratio of rotor exit tip diameter to rotor inlet diameter D_{3t}/D_2, to be limited to a maximum of 0·7 to avoid excessive shroud curvature (which could cause flow separation),

(e) the hub to tip diameter ratio at rotor exit to have a minimum value of 0·4 in order to avoid excessive blade blockage and total pressure loss at the hub.

Figure 8.10 shows the variation in total-to-static efficiency with specific speed (Ω_s) for a selection of nozzle exit flow angles, a_2. For each value of a_2 a hatched area is drawn, inside of which the various diameter ratios are varied. The envelope of maximum η_{ts} is bounded by the constraints $D_{3h}/D_{3t} = 0·4$ in all cases and $D_{3t}/D_2 = 0·7$ for $\Omega_s \geqslant 0·58$ in these hatched regions. This envelope is the *optimum geometry curve* and has a peak η_{ts} of 0·87 at $\Omega_s = 0·58$ rad. An interesting comparison is made by Rohlik with the experimental results obtained by Kofskey and Wasserbauer[21] on a single 90 deg IFR turbine rotor operated with several nozzle blade row configurations. The peak value of η_{ts}

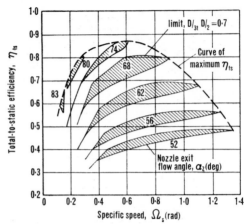

Fig. 8.10. Calculated performance of 90 deg IFR turbine (adapted from Rohlick[19]).

from this experimental investigation also turned out to be 0·87 at a slightly higher specific speed, $\Omega_s = 0·64$ rad.

The distribution of losses for optimum geometry over the specific speed range is shown in Fig. 8.11. The way the loss distributions change is a result of the changing ratio of flow to specific work. At low Ω_s all friction losses are relatively large because of the high ratios of surface area to flow area. At high Ω_s the high velocities at turbine exit cause the kinetic energy leaving loss to predominate. Figure 8.12 shows several meridional plane sections at three values of specific speed corresponding to the curve of maximum total-to-static efficiency. The ratio of nozzle exit height to rotor diameter, b_2/D_2, is shown in Fig. 8.13, the general rise of this ratio with increasing Ω_s reflecting the increase in nozzle flow area* accompanying the larger flow rates of higher specific speed. Figure 8.13 also shows the variation of U_2/c_0 with Ω_s along the curve of maximum total-to-static efficiency.

CLEARANCE AND WINDAGE LOSSES

A clearance gap must exist between the rotor vanes and the shroud. Because of the pressure difference between the pressure and suction

* The ratio b_2/D_2 is also affected by the pressure ratio and this has not been shown.

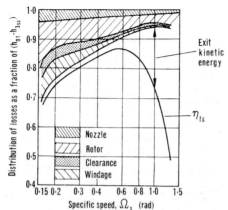

FIG. 8.11. Distribution of losses along envelope of maximum total-to-static efficiency (adapted from Rohlik[19]).

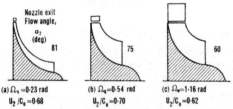

FIG. 8.12. Sections of radial turbines of maximum static efficiency (adapted from Rohlik[19]).

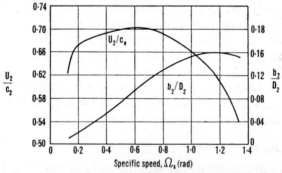

FIG. 8.13. Variation in blade speed/spouting velocity ratio (U_2/c_0) and nozzle blade height/rotor inlet diameter (b_2/D_2) corresponding to maximum total-to-static efficiency with specific speed (adapted from Rohlik[19]).

surfaces of a vane, a leakage flow occurs through the gap introducing a loss in efficiency of the turbine. The minimum clearance is usually a compromise between manufacturing difficulty and aerodynamic requirements. Often, the minimum clearance is determined by the differential expansion and cooling of components under *transient* operating conditions which can compromise the steady state operating condition. According to Rohlik[19] the loss in specific work as a result of gap leakage can be determined with the simple proportionality:

$$\Delta h_c = \Delta h_0(c/b_{av}) \qquad (8.35)$$

where Δh_0 is the turbine specific work uncorrected for clearance or windage losses and c/b_{av} is the ratio of the gap to average vane height (i.e. $b_{av} = \frac{1}{2}(b_2 + b_3)$). A constant axial and radial gap, $c = 0.25$ mm, was used in the analytical study of Rohlik quoted earlier. According to Rodgers,[9] extensive development on small gas turbines has shown that it is difficult to maintain clearances less than about 0.4 mm. One consequence of this is that as small gas turbines are made progressively smaller the *relative* magnitude of the clearance loss must increase.

The non-dimensional power loss due to windage on the back of the rotor has been given by Shepherd[1] in the form:

$$\Delta P_w/(\rho_2 \Omega^3 D_2{}^5) = \text{constant} \times Re^{-1/5}$$

where Ω is the rotational speed of the rotor and Re is a Reynolds number. Rohlik[19] used this expression to calculate the loss in specific work due to windage,

$$\Delta h_w = 0.56\rho_2 \ D_2{}^2(U_2/100)^3/(\dot{m} \ Re) \qquad (8.36)$$

where $\dot{m}$ is the total rate of mass flow entering the turbine and the Reynolds number is defined by $Re = U_2 D_2/\nu_2$, ν_2 being the kinematic viscosity of the gas corresponding to the static temperature T_2 at nozzle exit.

PRESSURE RATIO LIMITS OF THE 90 DEG IFR TURBINE

Every turbine type has pressure ratio limits, which are reached when the flow chokes. Choking usually occurs when the absolute flow at rotor exit reaches sonic velocity. (It can also occur when the *relative* velocity within the rotor reaches sonic conditions.) In the following analysis it is assumed that the turbine first chokes when the absolute exit velocity c_3 reaches the speed of sound. It is also assumed that c_3 is without swirl and that the fluid is a perfect gas.

For simplicity it is also assumed that the diffuser efficiency is 100% so that, referring to Fig. 8.4, $T_{04ss} = T_{03ss}$ ($p_{03} = p_{04}$). Thus, the turbine total-to-total efficiency is,

$$\eta_t = \frac{T_{01} - T_{03}}{T_{01} - T_{03ss}}. \tag{8.36}$$

The expression for the spouting velocity, now becomes

$$c_0^2 = 2C_p(T_{01} - T_{03ss}),$$

is substituted into eqn. (8.36) to give,

$$\eta_t = \frac{1}{1 - (T_{03ss}/T_{01})} - \frac{2C_p T_{03}}{c_0^2}. \tag{8.37}$$

The stagnation pressure ratio across the turbine stage is given by $p_{03}/p_{01} = (T_{03ss}/T_{01})^{\gamma/(\gamma-1)}$; substituting this into eqn. (8.37) and rearranging, the exhaust energy factor is,

$$\left(\frac{c_3}{c_0}\right)^2 = \left[\frac{1}{1 - (p_{03}/p_{01})^{(\gamma-1)/\gamma}} - \eta_t\right]\frac{c_3^2}{2C_p T_{03}}. \tag{8.38}$$

Now $T_{03} = T_3[1 + \frac{1}{2}(\gamma - 1)M_3^2]$ and

$$\frac{c_3^2}{2C_p} = T_{03} - T_{01} = T_3\left(\frac{\gamma - 1}{2}\right)M_3^2,$$

therefore,

$$\frac{c_3^2}{2C_p T_{03}} = \frac{\frac{1}{2}(\gamma - 1)M_3^2}{1 + \frac{1}{2}(\gamma - 1)M_3^2}. \tag{8.39}$$

With further manipulation of eqn. (8.38) and using eqn. (8.39) the stagnation pressure ratio is expressed explicitly as

$$\left(\frac{p_{01}}{p_{03}}\right)^{(\gamma-1)/\gamma} = \frac{(c_3/c_0)^2 + [\frac{1}{2}(\gamma - 1)M_3^2\eta_t]/[1 + \frac{1}{2}(\gamma - 1)M_3^2]}{(c_3/c_0)^2 - [\frac{1}{2}(\gamma - 1)M_3^2(1 - \eta_t)]/[1 + \frac{1}{2}(\gamma - 1)M_3^2]}. \tag{8.40}$$

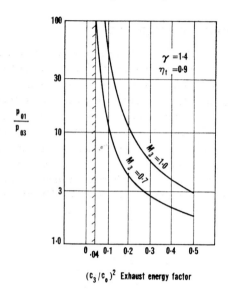

$(c_3/c_0)^2$ Exhaust energy factor

FIG. 8.14. Pressure ratio limit function for a turbine (Wood[18]) (By courtesy of the American Society of Mechanical Engineers.)

Wood[18] has calculated the pressure ratio (p_{01}/p_{03}) using this expression, with $\eta_t = 0.9$, $\gamma = 1.4$ and for $M_3 = 0.7$ and 1.0. The result is shown in Fig. 8.14. In practice, exhaust choking effectively occurs at nominal values of $M_3 \doteq 0.7$ (instead of at the ideal value of $M_3 = 1.0$) due to non-uniform exit flow.

The kinetic energy ratio $(c_3/c_0)^2$ has a first order effect on the pressure ratio limits of single stage turbines. The effect of any exhaust swirl present would be to lower the limits of choking pressure ratio.

It has been observed by Wood that high pressure ratios tend to compel the use of lower specific speeds. This assertion can be demonstrated by means of Fig. 8.8 taken together with Fig. 8.14. In Fig. 8.8, for a given value of A_3/A_d, Ω_s increases with $(c_3/c_0)^2$ increasing. From Fig. 8.14, (p_{01}/p_{03}) decreases with increasing values of $(c_3/c_0)^2$. Thus, for a given value of $(c_3/c_0)^2$, the specific speed must *decrease* as the design pressure ratio is increased.

COOLED 90 DEG IFR TURBINES

The incentive to use higher temperatures in the basic Brayton gas turbine cycle is well known and arises from a desire to increase cycle efficiency and specific work output. In all gas turbines designed for high efficiency a compromise is necessary between the turbine inlet temperature desired and the temperature which can be tolerated by the turbine materials used. This problem can be minimised by using an auxiliary supply of cooling air to lower the temperature of the highly stressed parts of the turbine exposed to the high temperature gas. Following the successful application of blade cooling techniques to axial flow turbines (see, for example, Horlock[22] or Fullagar[23]), methods of cooling small radial gas turbines have been developed.

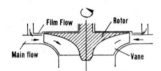

Fig. 8.15. Cross section of film-cooled radial turbine.

According to Rodgers,[9] the most practical method of cooling small radial turbines is by film (or veil) cooling, Fig. 8.15, where cooling air is impinged on to the rotor and vane tips. The main problem with this method of cooling being its relatively low *cooling effectiveness*, defined by

$$\varepsilon = \frac{T_{01} - (T_m + \Delta T_0)}{T_{01} - (T_{0c} + \Delta T_0)} \qquad (8.41)$$

where T_m is the rotor metal temperature,

$\Delta T_0 = \frac{1}{2} U_2^2 / C_p$, is half the drop in stagnation temperature of the gas as a result of doing work on the rotor,

T_{0c} is the stagnation temperature of the cooling air.

Rodgers refers to tests which indicate the possibility of obtaining $\varepsilon = 0.30$ at the rotor tip section with a cooling flow of approximately 10% of the main gas flow. Since the cool and hot streams rapidly mix, effectiveness decreases with distance from the point of impingement. A model study of the heat transfer aspects of film-cooled radial flow gas turbines is given by Metzger and Mitchell.[24]

REFERENCES

1. SHEPHERD, D. G. *Principles of turbomachinery.* Macmillan (1956).
2. KEARTON, W. J. *Steam turbine theory and practice* (6th edn.), Pitman, (1951).
3. PUYO, A. Hydraulic turbine development during the last few years. *La Houille Blanche,* 18 (1963).
4. SHMELEV, V. Bratsk hydroelectric project. *Water Power,* 21 (1969).
5. BAPTIST, J. V. and NITTA, R. Y. Large hydroelectric generators for the Grand Coulee third power plant. *Proc. Am. Power Conf.* 31 (1969).
6. DANEL, P. The hydraulic turbine in evolution. *Proc. Instn. Mech. Engrs. London,* 173 (1959).
7. ANON. Conceptual design study of a nuclear Brayton turboalternator–compressor. *Contractor Report, General Electric Company. NASA CR-113925* (1971).
8. BENSON, R. S., CARTWRIGHT, W. G. and DAS, S. K. An investigation of the losses in the rotor of a radial flow gas turbine at zero incidence under conditions of steady flow. *Proc. Instn. Mech. Engrs. London,* 182, Pt 3H (1968).
9. RODGERS, C. A cycle analysis technique for small gas turbines. "Technical Advances in Gas Turbine Design." *Proc. Instn. Mech. Engrs. London,* 183, Pt 3N (1969).
10. NUSBAUM, W. J. and KOFSKEY, M. G. Cold performance evaluation of 4·97 inch radial-inflow turbine designed for single-shaft Brayton cycle space-power system. *NASA TN D-5090*(1969).
11. BENSON, R. S. A review of methods for assessing loss coefficients in radial gas turbines. *Int. J. Mech. Sci.* 12 (1970).
12. FUTRAL, M. J. and WASSERBAUER, C. A. Off-design performance prediction with experimental verification for a radial-inflow turbine. *NASA TN D-2621* (1965).
13. BRIDLE, E. A. and BOULTER, R. A., A simple theory for the prediction of losses in rotors of inward radial flow turbines. *Proc. Instn. Mech. Engrs. London,* 182 Pt 3H (1968).
14. JAMIESON, A. W. H., The radial turbine. Ch. 9, *Gas turbine principles and practice,* ed. Sir H. Roxbee-Cox. Newnes (1955).
15. WALLACE, F. J., Theoretical assessment of the performance characteristics of inward radial flow turbines. *Proc. Instn. Mech. Engrs. London,* 172 (1958).
16. HIETT, G. F. and JOHNSTON, I. H., Experiments concerning the aerodynamic performance of inward radial flow turbines. *Proc. Instn. Mech. Engrs. London,* 178, Pt 3I (ii) (1964).
17. BALJE, O. E., A study on design criteria and matching of turbomachines: Part A, *J. of Eng. for Power. Trans. Am. Soc. Mech. Engrs.,* 84 (1962).
18. WOOD, H. J., Current technology of radial-inflow turbines for compressible fluids. J. of Eng. for Power. *Trans. Am. Soc. Mech. Engrs.,* 85 (1963).
19. ROHLIK, H. E., Analytical determination of radial-inflow turbine design geometry for maximum efficiency. *NASA TN D-4384* (1968).
20. STEWART, W. L., WHITNEY, W. J. and WONG, R. Y., A study of boundary-layer characteristics of turbomachine blade rows and their relation to overall blade loss. J. Basic Eng., *Trans. Am. Soc. Mech. Engrs.,* 82 (1960).

21. Kofskey, M. G. and Wasserbauer, C. A., Experimental performance evaluation of a radial inflow turbine over a range of specific speeds. *NASA TN D-3742* (1966).
22. Horlock, J. H. *Axial flow turbines*. Butterworths, London (1966).
23. Fullagar, K. P. L., The design of air cooled turbine rotor blades. *Symposium on Design and Calculation of Constructions Subject to High Temperature, University of Delft* (Sept., 1973).
24. Metzger, D. E. and Mitchell, J. W., Heat transfer from a shrouded rotating disc with film cooling. *J. of Heat Transfer, Trans. Am. Soc. Mech Engrs.*, **88** (1966).

PROBLEMS

1. A small inward radial flow gas turbine, comprising a ring of nozzle blades, a radial vaned impeller and an axial diffuser, operates at its design point with a total-to-total efficiency of 0·90. At turbine entry the stagnation pressure and temperature of the gas is 400 kPa and 1,140 K. The flow leaving the turbine is diffused to a pressure of 100 kPa and has negligible final velocity. Given that the flow is just choked at nozzle exit, determine the impeller peripheral speed and the flow outlet angle from the nozzles.

For the gas assume $\gamma = 1\cdot333$ and $R = 287$ J/(kg°C).

2. The mass flow rate of gas through the turbine given in Problem No. 1 is 3·1 kg/s, the ratio of the impeller axial width/impeller tip radius (b_2/r_2) is 0·1 and the nozzle isentropic velocity ratio (ϕ_2) is 0·96. Assuming that the space between nozzle exit and impeller entry is negligible and ignoring the effects of blade blockage, determine:

(i) the static pressure and static temperature at nozzle exit;
(ii) the impeller tip diameter and rotational speed;
(iii) the power transmitted assuming a mechanical efficiency of 93·5%.

3. A radial turbine is proposed as the gas expansion element of a nuclear powered Brayton cycle space power system. The pressure and temperature conditions through the stage at the design point are to be as follows:

Upstream of nozzles, $p_{01} = 699$ kPa, $T_{01} = 1,145$ K;
Nozzle exit, $p_2 = 527\cdot2$ kPa, $T_2 = 1,029$ K;
Rotor exit, $p_3 = 384\cdot7$ kPa, $T_3 = 914\cdot5$ K, $T_{03} = 924\cdot7$ K.

The ratio of rotor exit mean diameter to rotor inlet tip diameter is chosen as 0·49 and the required rotational speed as 24,000 rev/min. Assuming the relative flow at rotor inlet is radial and the absolute flow at rotor exit is axial, determine:

(i) the total-to-static efficiency of the turbine;
(ii) the rotor diameter;
(iii) the implied enthalpy loss coefficients for the nozzles and rotor row.

The gas employed in this cycle is a mixture of helium and xenon with a molecular weight of 39·94 and a ratio of specific heats of 5/3. The Universal gas constant is, $R_0 = 8\cdot314$ kJ/(kg-mol K).

4. A film-cooled radial inflow turbine is to be used in a high performance open Brayton cycle gas turbine. The rotor is made of a material able to withstand a temperature of 1145 K at a tip speed of 600 m/s for short periods of operation. Cooling air is supplied by the compressor which operates at a stagnation pressure ratio of 4 to 1, with an adiabatic efficiency of 80%, when air is admitted to the compressor at a stagnation temperature of 288 K. Assuming that the effectiveness of the film cooling is 0·30 and the cooling air temperature at turbine entry is the same as that at compressor exit, determine the maximum permissible gas temperature at entry to the turbine.

Take $\gamma = 1\cdot4$ for the air. Take $\gamma = 1\cdot333$ for the gas entering the turbine. Assume $R = 287$ J/(kg K) in both cases.

APPENDIX 1

Conversion of British Units to SI Units

Length
1 inch	$= 0.0254$ m
1 foot	$= 0.3048$ m

Force
1 lbf	$= 4.448$ N
1 ton f (UK)	$= 9.964$ kN

Area
1 in^2	$= 6.452 \times 10^{-4}$ m^2
1 ft^2	$= 0.09290$ m^2

Pressure
1 lbf/in^2	$= 6.895$ kPa
1 ft H$_2$O	$= 2.989$ kPa
1 in Hg	$= 3.386$ kPa
1 bar	$= 100.0$ kPa

Volume
1 in^3	$= 16.39$ cm^3
1 ft^3	$= 28.32$ dm^3
	$= 0.02832$ m^3
1 gall (UK)	$= 4.546$ dm^3

Energy
1 ft lbf	$= 1.356$ J
1 Btu	$= 1.055$ kJ

Velocity
1 ft/s	$= 0.3048$ m/s
1 mile/h	$= 0.447$ m/s

Specific energy
1 ft lbf/lb	$= 2.989$ J/kg
1 Btu/lb	$= 2.326$ kJ/kg

Mass
1 lb	$= 0.4536$ kg
1 ton (UK)	$= 1016$ kg

Specific heat capacity
1 ft lbf/(lb°F)	$= 5.38$ J/(kg°C)
1 ft lbf/(slug °F)	$= 0.167$ J/(kg°C)
1 Btu/(lb °F)	$= 4.188$ kJ/(kg °C)

Density
1 lb/ft^3	$= 16.02$ kg/m^3
1 slug/ft^3	$= 515.4$ kg/m^3

Power
1 hp	$= 0.7457$ kW

APPENDIX 2

Answers to Problems

Chapter 1

   1. $6·29 \text{ m}^3/\text{s}$.   **2.** $9·15 \text{ m/s}$; $5·33$ atmospheres.

   3. 551 rev/min, 1:10·8; $0·885 \text{ m}^3/\text{s}$; 17·85 MN.

   4. 4,030 rev/min; 31·4 kg/s.

Chapter 2

   1. 88·1 per cent.   **2.** (i) 704 K; (ii) 750 K; (iii) 668 K.

   3. (i) 500 K, $0·313 \text{ m}^3/\text{kg}$; (ii) 1·042.   **4.** 49·1 kg/s; 24 mm.

   5. (i) 630 kPa, 275°C; 240 kPa, 201°C; 85 kPa, 126°C; 26 kPa, $q = 0·988$; 7 kPa, $q = 0·95$; (ii) 0·638, 0·655, 0·688, 0·726, 0·739; (iii) 0·739, 0·724; (iv) 1·075.

Chapter 3

   1. 49·8 deg.   **2.** 0·77; $C_D = 0·048$, $C_L = 2·15$.   **3.** $-1·3$ deg, 9·5 deg, 1·11.

   4. (i) 53 deg and 29·5 deg; (ii) 0·962; (iii) $2·17 \text{ kN/m}^2$.

   5. (a) $s/l = 1·0$, $a_2' = 24·8$ deg; (b) $C_L = 0·872$.

   6. (b) 57·8 deg; (c) (i) 357 kPa; (ii) 0·96; (iii) 0·0218, 1·075.

Chapter 4

   2. 22·7 kJ/kg; 420 kPa, 177°C.   **3.** 91%.

   4. (i) 1·503; (ii) 39·9 deg, 59 deg; (iii) 0·25; (iv) 90·5 and 81·6%.

   5. $c_x = 148·4 \text{ m/s}$, $R = 0·152$, $\eta_{tt} = 89·9\%$.

   6. (i) 215 m/s; (ii) 0·098, 2·68; (iii) 0·872; (iv) 265°C, 0·75 MPa.

Chapter 5

   1. 14 stages. **2.** 30·6°C. **3.** 132·5 m/s, 56·1 kg/s; 10·1 MW.

   4. 86·5 %; 9·28 MW.   **5.** 0·59, 0·415.

   6. 33·5 deg, 8·5 deg, 52·9 deg; 0·827; 34·5 deg, 1·11.

   7. 56·9 deg, 41 deg; 21·8 deg.

Chapter 6

   1. 55 and 47 deg.   **2.** 0·602, 1·38, $-0·08$ (implies large losses near hub)

   4. 70·7 m/s

5. Work done is constant at all radii;

$c_{x1}{}^2 = \text{constant} - 2a^2[(r^2-1)-2(b/a)\ln r);$

$c_{x2}{}^2 = \text{constant} - 2a^2[(r^2-1)+2(b/a)\ln r];$

$\beta_1 = 43{\cdot}2$ deg, $\beta_2 = 10{\cdot}4$ deg.

6. (i) 480 m/s; (ii) 0·818; (iii) 0·08; (iv) 3·42 MW; (v) 906·8K, 892·2K.

7. (i) 62 deg; (ii) 61·3 and 7·6 deg; (iii) 45·2 and 55·9 deg, (iv) −0·175, 0·477.

8. See Fig. 6.13. For (i) at $x/r_t = 0{\cdot}05$, $c_x = 113{\cdot}2$ m/s.

Chapter 7

1. 579 kW; 169 mm; 5·0.

2. 0·875; 5·61 kg/s.

3. 26,800 rev/min; 0·203 m, 0·525.

4. 0·735, 90·5%.

5. (i) 542·5 kW; (ii) 536 and 519 kPa; (iii) 586 and 240·8 kPa, 1·20, 176 m/s; (iv) 0·875; (v) 0·22; (vi) 28,400 rev/min.

6. (i) 29·4 dm³/s; (ii) 0·781; (iii) 77·7 deg; (iv) 7·82 kW.

Chapter 8

1. 586 m/s, 73·75deg. **2.** (i) 205·8 kPa, 977 K; (ii) 125·4 mm, 89,200 rev/min; (iii) 1000 kW.

3. (i) 90·3%; (ii) 269 mm; (iii) 0·051, 0·223. **4.** 1593 K.

Index

5000000 00 00 0 00
0000 0000 0 00 0
pages